AF669956

EVENT-RELATED POTENTIALS AND EVOKED POTENTIALS

Edited by **Phakkharawat Sittiprapaporn**

Event-Related Potentials and Evoked Potentials
http://dx.doi.org/10.5772/65183
Edited by Phakkharawat Sittiprapaporn

Contributors

Manuel Vazquez-Marrufo, Reza Fazel-Rezai, Ali Haider, Stefan Grossauer, Katharina Koeck, Yaroslaw Parpaley, Helene Selpien, Daniel Dumalin, Silvio Pessanha Neto, Regina Maria Papais Alvarenga, Luiz Carlos Pinto

Published by InTech

Janeza Trdine 9, 51000 Rijeka, Croatia

© The Editor(s) and the Author(s) 2017
The moral rights of the editor(s) and the author(s) have been asserted.
All rights to the book as a whole are reserved by InTech. The book as a whole (compilation) cannot be reproduced, distributed or used for commercial or non-commercial purposes without InTech's written permission. Enquiries concerning the use of the book should be directed to InTech's rights and permissions department (permissions@intechopen.com).
Violations are liable to prosecution under the governing Copyright Law.

(cc) BY

Individual chapters of this publication are distributed under the terms of the Creative Commons Attribution 3.0 Unported License which permits commercial use, distribution and reproduction of the individual chapters, provided the original author(s) and source publication are appropriately acknowledged. More details and guidelines concerning content reuse and adaptation can be found at http://www.intechopen.com/copyright-policy.html.

Notice

Statements and opinions expressed in the chapters are these of the individual contributors and not necessarily those of the editors or publisher. No responsibility is accepted for the accuracy of information contained in the published chapters. The publisher assumes no responsibility for any damage or injury to persons or property arising out of the use of any materials, instructions, methods or ideas contained in the book.

Publishing Process Manager Nina Kalinic
Technical Editor SPi Global
Cover InTech Design team

First published December, 2017
Printed in Croatia
Legal deposit, Croatia: National and University Library in Zagreb

Additional hard copies can be obtained from orders@intechopen.com

Event-Related Potentials and Evoked Potentials, Edited by Phakkharawat Sittiprapaporn
p. cm.
Print ISBN 978-953-51-3639-2
Online ISBN 978-953-51-3640-8

We are IntechOpen,
the world's leading publisher of
Open Access books
Built by scientists, for scientists

3,650+
Open access books available

114,000+
International authors and editors

118M+
Downloads

151
Countries delivered to

Our authors are among the
Top 1%
most cited scientists

12.2%
Contributors from top 500 universities

WEB OF SCIENCE™

Selection of our books indexed in the Book Citation Index
in Web of Science™ Core Collection (BKCI)

Interested in publishing with us?
Contact book.department@intechopen.com

Numbers displayed above are based on latest data collected.
For more information visit www.intechopen.com

For EU product safety concerns:
IN TECH d.o.o., Prolaz Marije Krucifikse Kozulić 3, 51000 Rijeka, Croatia,
info@intechopen.com or visit our website at intechopen.com.

Meet the editor

Dr. Wichian Sittiprapaporn is a Thai cognitive neuroscientist who has published in many national and international publications. His other achievements include authoring chapters in other books and reviewing articles for reputed national and international indexed journals. He was awarded Scholar of the Month for July 2009 for his interdisciplinary research and educational leadership awarded by the Guild of Independent Scholars, Journal of Alternative Perspectives in the Social Sciences. Dr. Wichian Sittiprapaporn was born in Udon Thani, Thailand, in 1970. He received his BA (Hons) (English) and MA (Linguistics) degrees from Srinakharinwirot University and Mahidol University, Bangkok, Thailand, in 1993 and 1997, respectively. He received a PhD degree in Neurosciences from the Institute of Science and Technology for Research and Development, Mahidol University, Bangkok, Thailand, in 2002, and became a postdoctoral fellow at the Clinical Cognitive Neuroscience Center (CCNC), Seoul National University, College of Medicine, Seoul, Korea, in 2006. After working as a lecturer in Neurosciences at the Neuro-Behavioural Biology Center, Institute of Science and Technology for Research and Development, Mahidol University, from 2003 to 2005, he was invited to join the Music Therapy Division, College of Music, Mahidol University, Bangkok, in 2008. In January 2010, he joined the College of Music, Mahasarakham University, Mahasarakham, and concurrently held a lecturer position in Faculty of Medicine Mahasarakham University and College of Research Methodology and Cognitive Science, Burapha University, Thailand. His current research includes cognitive neurosciences and biomedical image and signal processing of human brain function in music perception and cognition.

Contents

Preface

This edited volume is a collection of reviewed and relevant research chapters, concerning the developments within the event-related potentials and evoked potentials field of study. An event-related potential (ERP) is the measured brain response that is the direct result of a specific sensory, cognitive, or motor event. More formally, it is any stereotyped electrophysiological response to a stimulus. The study of the brain in this way provides a noninvasive means of evaluating brain functioning. ERPs are measured by means of electroencephalography (EEG).

The book includes scholarly contributions by various authors and is edited by a group of experts pertinent to medicine and health sciences. Each contribution comes as a separate chapter complete in itself but directly related to the book's topics and objectives.

This book includes chapters dealing with the topics: "Event-Related Potentials for the Study of Cognition," "Application of P300 Event-Related Potential in Brain-Computer Interface," "Motor-Evoked Potentials in Supratentorial Glioma Surgery," "The Contingent Negative Variation: The Cumulative Curve Method Revisited," and "The Visual-Evoked Potential in Idiopathic Inflammatory Demyelinating Diseases."

The target audience comprises scholars and specialists in the field.

InTechOpen

Chapter 1

Event-Related Potentials for the Study of Cognition

Manuel Vazquez-Marrufo

Additional information is available at the end of the chapter

http://dx.doi.org/10.5772/intechopen.69308

Abstract

Despite the vast literature on event-related potentials (ERPs), many clinical professionals are still unaware of the huge variety of possible applications they offer. The aim of this chapter is not to show the classical use of ERPs, focused on analyzing the first steps of information processing (sensory pathways). On the contrary, this chapter will be focused on the use of these ERPs in the assessment of cognitive function. In particular, this chapter is mainly focused on the use of ERPs to better understand the neural bases of cognitive impairment from the electrical activity of the brain. Describing all the possible ERP components and their cognitive meaning is a huge endeavor, and this chapter will only be focused on three of them: contingent negative variation (CNV), mismatch negativity (MMN), and P300. To improve the reader's knowledge about these ERPs in cognition, a specific description will be given about the stimulation required to obtain the specific component, the topography, and latency shown. Moreover, a description of the neurophysiological bases of the component, its relationship with psychological processes and neural sources will be also included. Pathological alterations suffered by the component will also be briefly described.

Keywords: cognition, ERPs, latency, neural sources, pathology, topography

1. Introduction

Since the 1960s, a prolific literature has been produced on the field of event-related potentials (ERPs), related to the study of cognitive activity in the brain. In the beginning, these studies were more directed to the study of sensory and motor pathways. However, from studies such as in Refs. [1–3], ERPs were related to cognitive processes such as relevance of the stimulus, uncertainty, or mismatch with a previous stimulus.

Up to the present day, many studies have been published by numerous groups worldwide using this technology. In spite of the crisis that the ERP technique suffered due to the arrival

INTECH
open science | open minds

© 2017 The Author(s). Licensee InTech. This chapter is distributed under the terms of the Creative Commons Attribution License (http://creativecommons.org/licenses/by/3.0), which permits unrestricted use, distribution, and reproduction in any medium, provided the original work is properly cited. (cc) BY

of neuroimaging, ERPs have survived, and nowadays, still offer an interesting way to explore the cognitive activity with a direct measure of the electrical activity of neurons.

One of the main challenges that ERPs have to overcome is their application in the clinical field through studies that verify the reliability of the technique. Nevertheless, it is necessary to define the possible applications of these potentials to better understand the etiology of cognitive pathology and develop possible therapeutic targeting in this field.

In this chapter, three important ERPs will be detailed: contingent negative variation (CNV), mismatch negativity (MMN), and P300. For each one, several topics will be tackled: (1) a generic description of the component; (2) a brief definition of a procedure that allows evoking the component; (3) evidence of psychological variables that can modulate the component; (4) neurophysiological basis, typical topography, and latency of the component; (5) neural sources identified; (6) alterations that the component suffers in some diseases and its probable meaning.

These three components have been selected because of the order in which they appear in information processing. The first one, CNV, is related to the instants prior to the onset of a stimulus that is expected by the subject. MMN is related to the early phases of the cognitive processing for the stimulus that is evaluated. And finally, P300 represents late phases of the perceptual process and includes many psychological processes of high order.

2. Contingent negative variation (CNV)

2.1. Generic description

In 1964, Walter et al. [2] published a study in which an event-related potential was present prior to the appearance of the stimuli. The psychological meaning of this component was defined as the expectancy caused by a warning stimulus (also called sometimes "cue") that allows the subject to prepare a response in order to react faster and more accurately to the incoming stimuli (known as "imperative stimuli").

2.2. Procedure and characteristics of the component

Diverse paradigms have been used to elicit CNV in diverse sensory modalities: visual [4]; auditory [5]; or the interaction between visual and auditory stimulation [6], and even with proprioceptive information [7]. In the last few years, one of these paradigms has been called "Attentional Network Test (ANT)," which has been highly popular in the study of attentional mechanisms such as expectancy or orienting [8–10]. Depending on the stimulus onset asynchrony (SOAs) used, CNV is present between warning and imperative stimuli in this task [4, 11]. In a basic conception of the ANT paradigm, cues are shown for 150 ms, and then a variable SOA can be defined in a range between 1000 and 2000 ms when the CNV is present. Finally, an imperative stimulus is displayed, and the subject has to respond according to the instructions of the task (see Ref. [4] for a complete description of the parameters of the task; see also **Figure 1** for a schematic of this procedure).

http://dx.doi.org/10.5772/intechopen.69308

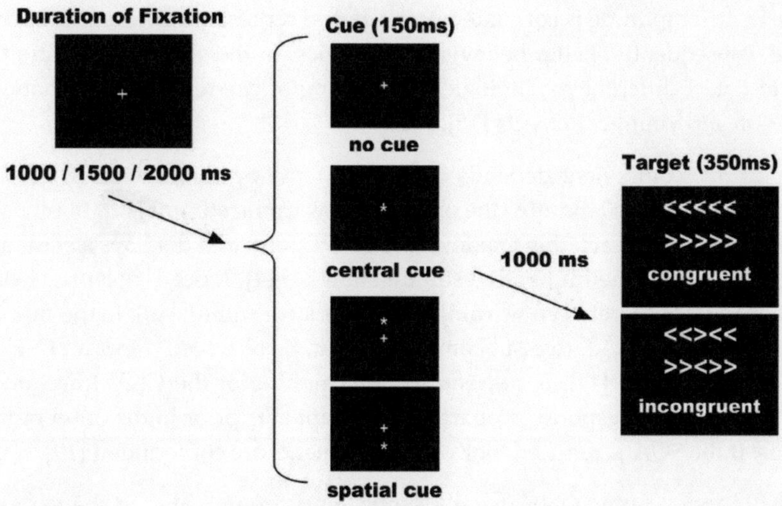

Figure 1. Schematic representation of attention network test. Adapted from Galvao-Carmona [4].

During its history, CNV has been studied extensively in terms of psychological variables that can modulate it. One of the earliest studies was about how uncertainty affects CNV [12]. In the case that the subject is not certainly sure when the imperative stimuli will be displayed, the amplitude grows fast; however, in the case that the subject knows approximately when the imperative stimuli will be presented, the amplitude grows gradually. In the case that there is no need to respond to the stimuli following the warning cue, CNV is not usually elicited [12]. However, some studies have shown that a nonmotoric activity is evoked in the absence of direct overt motor activity [13].

Another interesting fact is that even when the subject is not warned by a cue, there is a slow negative trend in the human brain that represents our general expectancy during an experimental session [4] (see **Figure 2**).

In regard to its relationship with development, Segalowitz and Davies [14] published a study in which it is possible to see the evolution of this component along infancy and adolescence.

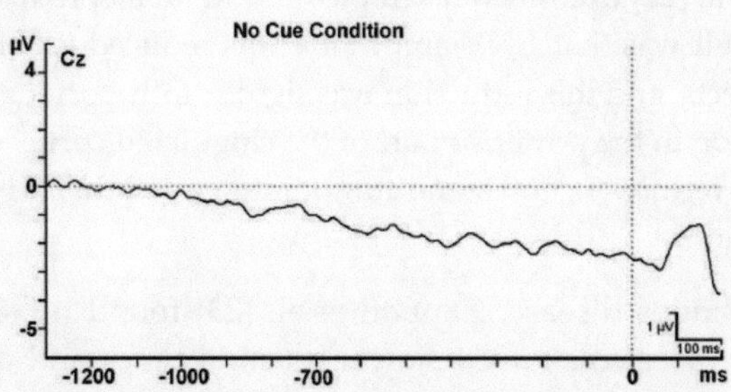

Figure 2. CNV component modulation in the no-cue condition showing a general expectancy during the execution of a warning-target paradigm. Zero value in *x*-axis represents the onset of the imperative stimuli. Adapted from Galvao-Carmona [4].

An increase in the amplitude is correlated with age and represents the maturation of the frontal lobe and, consequently, better behavioral capacities. In the elderly population, CNV has been used to detect different psychological processing of cue-relevant information between this population and younger subjects [15].

The latency of this component depends on the task employed. Sometimes, CNV reaches a peak (or valley) around 400 ms after the onset of the warning stimuli [12]. In other occasions, the component does not reach this maximum negative point and displays a continuous trend to negative values until the imperative stimuli show up [4]. Indeed, one of the critical variables for CNV is the SOA between warning and imperative stimuli [16]. In the case that a SOA of 3 or more seconds is used, two subcomponents can be observed. First, an O-wave, where O represents "Orienting" [17], is present at the beginning of the CNV trace, and then, an E-wave (expectancy and response preparation) [18] appears prior to the onset of the imperative stimulus. If the SOA is reduced, both subcomponents are confounded [19].

With respect to topography, CNV usually shows a maximum value in the vertex, which is symmetrically distributed over the scalp [4]. However, if the subcomponents are clearly distinguished, the O-wave is mainly frontal, and the E-wave is more postcentrally located [19].

The identification of the neural sources for this component remains under debate, perhaps due to the complete set of different processes present in the CNV latencies. Using magnetoencephalography, some authors [20] determined that the neural source for the magnetic counterpart of the CNV was located in the premotor cortex (Brodmann Area 6). In another study, Zappoli described that patients with lobotomy of frontal lobes exhibited decreased amplitude of the CNV [21]. In a study performed in our lab, in which different time intervals of the CNV trace were analyzed, numerous cortical areas were active, and a complex dynamic was present during the process [4]. These cortical areas belong to different lobes, including the frontal, parietal, occipital, and other regions, such as the cingulated lobe and insula, among others (see **Figure 3**).

2.3. Psychological meaning and pathology

Once the component was described, many studies have been performed to find alterations in the component and the possible meaning in diverse pathologies. In Huntington disease (HD), de Tommaso et al. [22] examined a sample of mild, demented, and nonmedicated HD patients. The main result was that CNV amplitude was reduced in these patients compared to healthy control subjects, and this reduction was significantly correlated to the bradykinesia score. A strong activation in the posterior part of the cingulated cortex in HD is likely responsible for the amplitude reduction, and some authors suggested that it is probably caused by a basal ganglia dysfunction.

With regard to Alzheimer's disease, Zappoli et al. [23] found no significant CNV activity in these patients, who also showed slower reaction times and other EEG alterations. However, another group [24] observed that the CNV amplitude was not different between groups, also showing low test-retest reliability, which makes it difficult to be applied in the clinical field.

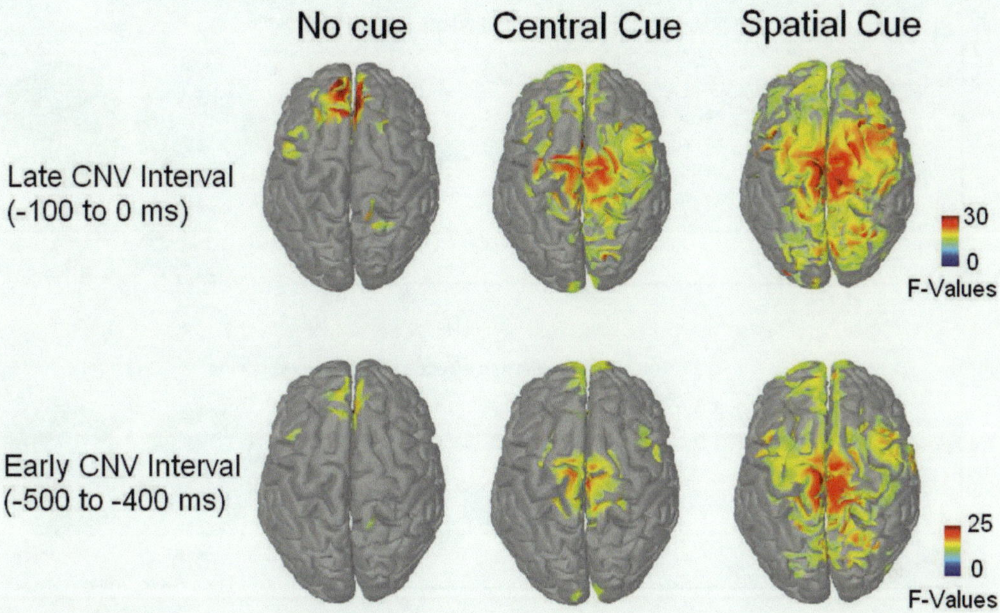

Figure 3. Cortical activation maps presented in Z-scores according to the baseline and showing significant activity (FDR-adjusted $p < 0.01$). Sources of the CNV effect were estimated in 2 CNV intervals of interest, −500 to −400 and −100 to 0 ms before the target stimulus. Adapted from Galvao-Carmona [4].

In other pathologies, CNV has been used to determine if any anatomical structure could be related to a specific cognitive impairment. For instance, Kuoppamäki et al. [25] observed that Parkinson patients with bilateral lesions in the globus pallidus present a deficit in motor tasks and alterations in the early phases of CNV.

From our laboratory, a study in a sample of multiple sclerosis patients, reduced amplitude of CNV was associated with impairment in the alerting and orienting attentional mechanisms. These results were also in accordance with neuropsychological scores from attentional tests [26] (see **Figure 4**).

In the psychopathological field, one of the main disorders studied with CNV has been schizophrenia. Some authors have described a reduction of the amplitude related to the frontal lobe dysfunction, and it was manifested in the frontal-central derivations and at the early CNV phase [27]. At the same time, some studies have been focused on the relationship between CNV and some items of questionnaires used in the assessment of negative or positive symptoms [28]. Another interesting field is related to the study of the neural mechanisms underlying cephalea and migraine. In their study, Siniatchkin et al. [29] selected three groups: migraine, chronic daily headache, and healthy control subjects. CNV values were lower for the migraine group, especially at the beginning of the CNV. Chronic daily headache patients showed a reduced negativity of the late component of CNV. An interesting result was the absence of habituation to CNV in both types of patients and the potential application of CNV in diagnostic and therapeutic strategies for these pathologies.

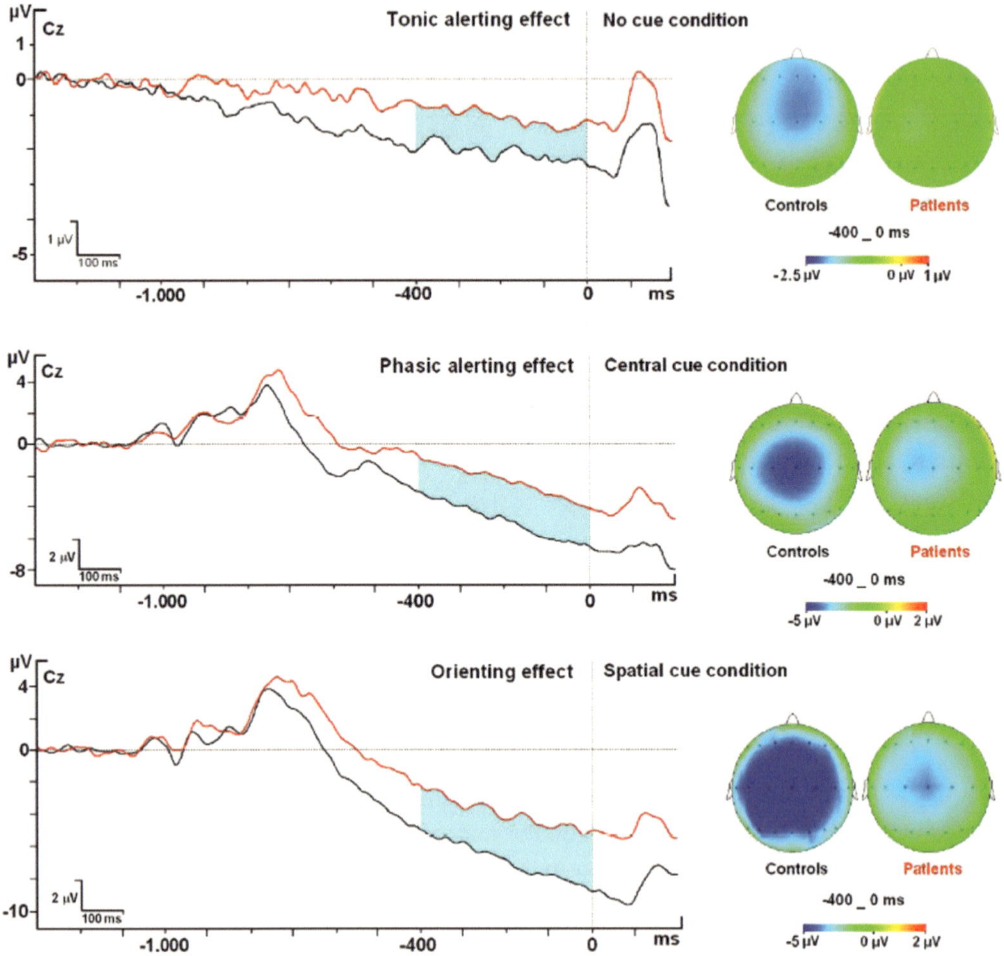

Figure 4. Contingent negative variation modulations at Cz electrode and topographic maps for healthy control subjects and patients in the attention network test. Adapted from Vazquez-Marrufo [26].

3. Mismatch negativity (MMN)

3.1. Generic description

Described for the first time by Näätänen et al. [3], mismatch negativity appears when a change in a stream of stimulation is detected. This generic fact has been employed in different approaches for studying the bases of cognition in healthy and pathological subjects.

3.2. Procedure and characteristics of the component

A typical way to obtain this component consists of using an auditory oddball task, in which two types of stimuli are listened to binaurally through headphones: standard stimuli (1000 Hz tones and a probability of 0.80) and deviant stimuli (2000 Hz tones and a probability of 0.20).

The interstimulus interval can be around 1 s, and the intensity of auditory stimuli can be set at 70 dB. The duration of the stimuli is 50-ms plateau and a 10-ms rise-fall time. Two blocks with 200 trials (including 80 deviant stimuli) are enough to obtain the MMN [30]. See **Figure 5** for a schematic representation of the experimental procedure.

The component can be elicited during active tasks (counting deviant stimuli) [30] or during passive tasks [3]. Indeed, this last option can be extremely useful in some pathological conditions, such as coma [31]. MMN shows up as the result of subtraction between the standard and deviant associated waves. This component is evoked not only by a change in the frequency but also in pitch duration, intensity of stimuli…, and so forth. [32]. Other properties such as short SOAs [33] or the saliency of the deviant stimuli [34] produce greater MMN amplitudes.

Although MMN is usually based on auditory procedures, it can also be obtained with visual stimulation [35, 36] or even other sensory modalities [37]. The component is present even in newborns [38], and, during childhood, MMN presents differences in latency and topography with respect to adults, which suggests a development of the component throughout youth [39]. In healthy aging, elderly subjects showed a reduction in the amplitude [40], as well as a delay in the latency [41].

With regard to the specific parameters of the component, the latency is between 150 and 250 ms after the onset of the stimuli, and its distribution is fronto-central in the scalp (although the topography depends on the location of the reference) [42]. In auditory paradigms, neural generators are located in the primary and nonprimary auditory cortex, although they can also include frontal lobe areas, the thalamus, and the hippocampus, as evidenced by intracranial studies with animals [42].

3.3. Psychological meaning and pathology

The main application of MMN has probably been as an exponent of accuracy in the discrimination of small changes in stimuli in untrained [43] or trained subjects [44]. Since the presence of standard stimuli is necessary for obtaining MMN in deviant stimuli, this component has

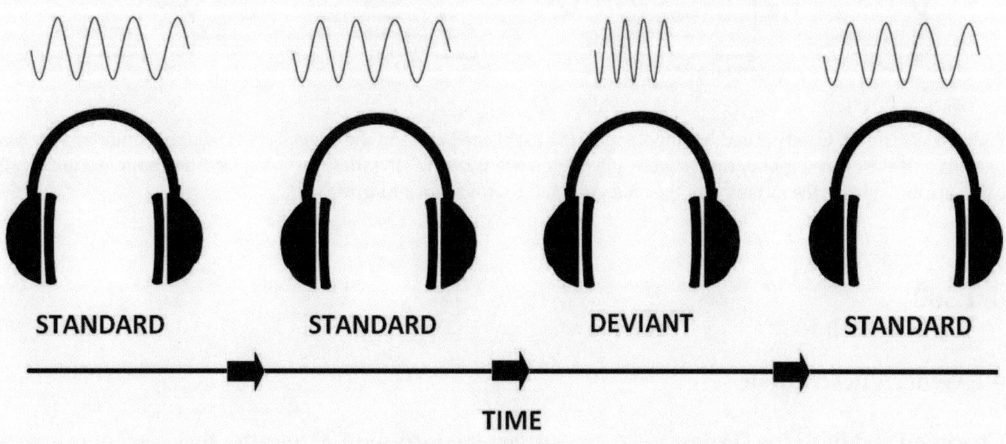

Figure 5. Schematic representation of the experimental procedure to evoke a MMN response.

also been proposed as an index of the violation of the memory trace built during the experiment by the standard stimuli [45].

With regard to its application in several pathologies, attenuated amplitudes in patients with schizophrenia have been reported in Ref. [46]. This reduction has been interpreted as a poor social/occupational and executive functioning in these patients [47].

With respect to bipolar disorder, several studies have shown contradictory results for this component (for a review, see Ref. [45]. The main conclusion is that there is no clear evidence of auditory discrimination ability in these patients after all.

In multiple sclerosis, some studies have reported alterations in this component, showing deficits in the auditory discrimination system [48]. Moreover, some authors have shown that the amplitude reduction in MMN could be linked to disorganization of spectral modulations (beta and gamma bands) in patients with low EDSS. These results suggest a complex set of alterations even in the early phases of this disease [30] (see **Figure 6**).

In stroke patients, an amplitude reduction in MMN has been found for changes in tone duration and frequency after a left-hemisphere stroke [49]. Another approach in the stroke field has been the use of this component as an assessment of function recovery in patients [50].

With regard to development, some studies have been focused on the use of MMN to determine deficits in dyslexic children. In particular, a reduction in the amplitude of this component found by Shafer et al. [51] could represent a poor auditory discrimination or language learning disability for phonetic cues in these patients. In autism spectrum disorder, some studies have shown an increase in the amplitude of this component with nonspeech stimuli and the opposite effect with speech stimuli [52, 53].

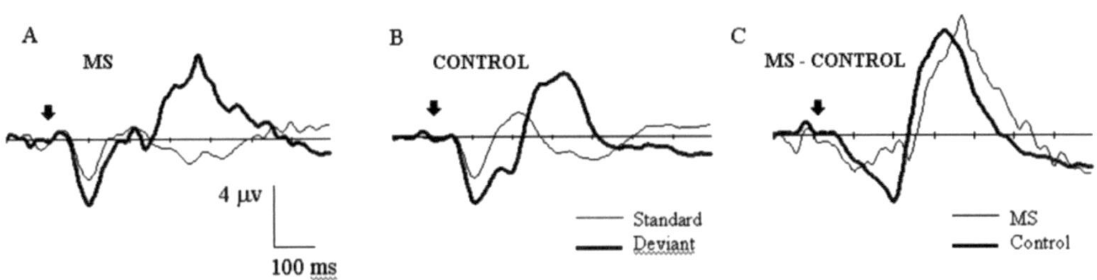

Figure 6. (A and B). Event-related brain potentials (ERPs) elicited at Cz in the deviant and standard conditions for MS patients and the control group, respectively. (C). Difference wave (MMN) (deviant—standard) for both groups. In all cases, arrows indicate the incoming of the tone. Adapted from Vazquez-Marrufo [30].

4. P300

4.1. Generic description

Chapman and Bragdon [1] described a positive wave around 300 ms after the onset of numerical and nonnumerical visual stimuli, and the subject was required to solve a problem with those numbers. These authors suggest that this positive wave was originated because the numbers were relevant to the task. A vast number of studies have found multiple possible meanings for this component, and a general consensus accepts that P3 represents the summation of different areas in the brain with diverse psychological processes intertwined. Indeed, P3 has two clear distinguishable subcomponents with different psychological meanings. In a simplified conception, P3b is evoked by relevant stimuli (target) and not usually evoked by standard stimuli in several paradigms (i.e., oddball task). On the other hand, P3a is evoked by the presence of novel stimuli along a sequence of target and standard stimuli.

4.2. Procedure and characteristics of the component

In this section, a brief description of a visual oddball paradigm is presented in comparison to the auditory type described in the MMN section (see Ref. [54] for complete specifications). In this "visual oddball task," the subject is asked to discriminate uncommon visual stimuli (target) from a sequence of frequent stimuli (standard). In this study, the target stimulus (probability: 25%) was a rectangle with a checkerboard pattern comprising red and white squares. The standard stimulus (probability: 75%) was equivalent in size and pattern but with black and white squares. Both stimuli were presented in the center of the screen and the size of both stimuli was 7.98 and 9.42 (visual angle) on the x and y axes, respectively. The duration of the stimuli was 500 ms, and the interstimulus interval was 1 s, which is the time when the subject could respond. The task for the participants was to press a button whenever a target stimulus appeared and ignore the standard stimuli. It is also possible to elicit P3 if the task is not a motor response, e.g., the subject just counts the targets silently [55]. Only one block with 200 trials (50 target stimuli) is enough to evoke the P3 component (see **Figure 7**).

Multiple studies have defined variables that can modulate this component. An interesting finding is about P3 being evoked by the absence of a stimulus if it is relevant to the subject [56]. Another important issue about this component is that it has been observed with different

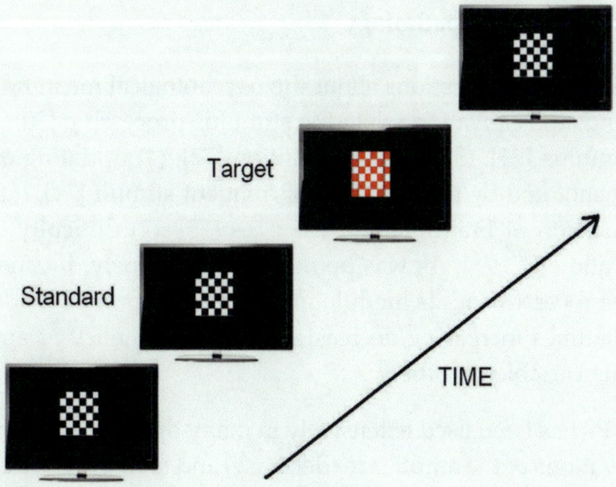

Figure 7. Schematic representation of a visual oddball to elicit P3.

sensory modalities (auditory, visual, somatosensory, olfactory, or taste stimulation) [57]. P3 amplitude is mainly independent of sensory modality; however, it is possible to find some differences in shape and latency when auditory and visual stimulation are compared [58]. When using auditory stimulation, different features of the stimulation, such as the tone frequency or the use of a mask of white noise, can affect P3 latency [59].

An important consensus regarding the meaning of P3 is that it reflects the timing of cognitive processes. However, on the other hand, it is not correlated with reaction time [60]. Considering the multiple psychological processes comprised in the component, there does not seem to exist a strong correlation between P3 and behavioral response.

With regard to age, children showed an increase in latency and a decrease in amplitude in the 1st years of life compared to adults (up to 3 years) [61]. Considering the entire lifespan, Goodin et al. [62] showed the natural evolution of latency decrease and amplitude increase in young subjects and that of latency increase and amplitude decrease in elderly subjects. A considerable number of publications have been focused on studying this component in other species. This component or similar waves have been described in rats [63], cats [64], monkeys [65], or mice [66], among other studies.

P3 latency peaks around 350 ms and, in particular, P3a and P3b are around 240 ms and 350 ms, respectively [67]. However, it is possible to find a P3 peak in a range that goes from 300 to 500 ms depending on many variables (type of task, difficulty…, and so forth). [55]. With regard to topography, the maximum amplitude of the P3 wave is seen at the parieto-occipital area for P3b and as fronto-central derivations for P3a [57]. With aging, the topography can change with a more frontal distribution; however, the scalp distribution is defined similarly by task requirements when it is compared with young subjects [68]. Concerning neural sources for this component, multiple studies have described controversial results about them. In particular, diverse cortical lobes (frontal, parietal, and temporal) or the hippocampus are defined as relevant for the generation of the P3 component (see Ref. [69] for a review).

4.3. Psychological meaning and pathology

Nowadays, there are many suggestions about the psychological meaning of this component: (1) inhibition that ends the activation related to stimulus processing [70]; (2) expectation and relevancy of the stimulus [71]; (3) selective attention [72]; (4) updating of working memory [73]; (5) activation generated by the sequence of frequent stimuli [74]; (6) speed of cognitive processing and allocation of brain energy resources [75]; (7) difficulty of the task [76]; (8) emotion and motivation [77, 78]. As was pointed out previously, it can be asserted that P3 comprises multiple processes and its modulation can be determined by different variables in different ways, sometimes increasing/decreasing either the latency or amplitude and sometimes opposing some variables to others.

In the clinical field, P3 has been used extensively in many diseases. Our group has referred in some studies to alterations of the amplitude (decrease) and latency (increase) of P3 in multiple sclerosis [79, 80] (see **Figure 8**). Comi et al. [81] has shown that a longer latency in P3 may be related to demyelination. An increased latency is also observed in diverse types of dementia

http://dx.doi.org/10.5772/intechopen.69308

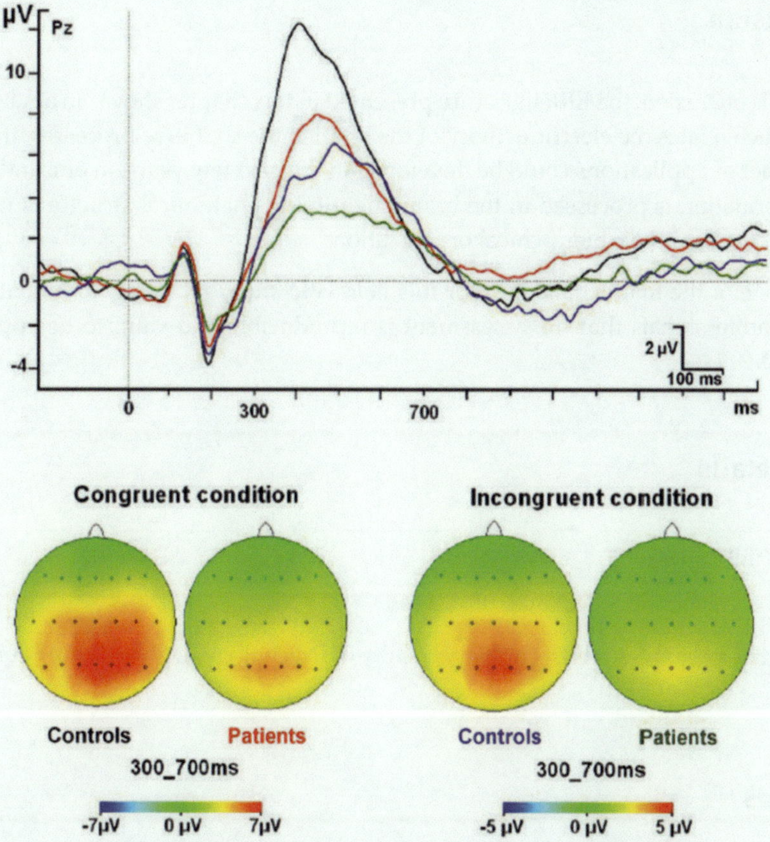

Figure 8. P3 component modulations at Pz electrode and topographic maps for ANT test. Note the reduction in the amplitude for multiple sclerosis patients in both conditions (congruent and incongruent). Adapted from Vazquez-Marrufo [80].

(Alzheimer, multiinfarction dementia, and lacunar dementia); on the other hand, in pseudodementia, the altered parameter is amplitude, which is flattened [82].

In Parkinson's disease, an amplitude decrease has also been observed by O'Donnel et al. [83]. However, other authors suggest that this reduction is more related to the dementia associated with the disease, rather than Parkinson itself [84].

P3 has also been useful as an indicator of the presence of a traumatic lesion (e.g., prefrontal cortex). It has been related to P3a and behavioral responses indicating a reduced attentional shift toward novel stimuli [85]. It is also possible to assess the evolution in the subacute phase of a stroke from changes in the P3 component [86].

In the psychopathology field, schizophrenia has received a remarkable attention with P3 studies. One general finding is decreased amplitude, which seems to be correlated to the presence of negative symptoms [87, 88]. Another potential application of P3 consists of assessing the neurodegenerative process in this pathology. Martin-Loeches showed a negative correlation between P3 amplitude and prefrontal CSF volume in these patients [89].

5. Conclusion

As a general conclusion, the ERP literature presented in this chapter shows an amazing field to explore, which relates the electric activity of the brain to the cognitive processes. It seems that a vast number of applications could be developed in the next few years, in our understanding of how information is processed in the brain, identifying anatomical structures where these processes occur, and their hierarchical organization.

However, one of the main challenges for this field is to study reliability tests that guarantee the health professionals that the assessment is reproducible and valid to be applied in the clinical field.

Author details

Manuel Vazquez-Marrufo

Address all correspondence to: marrufo@us.es

Department of Experimental Psychology, Faculty of Psychology, University of Seville, Seville, Spain

References

[1] Chapman RM & Bragdon HR. Evoked responses to numerical and non-numerical visual stimuli while problem solving. Nature. 1964;**203**:1155-1157. DOI: 10.1038/2031155a0

[2] Walter WG, Cooper R, Aldridge VJ, McCallum WC, Winter AL. Contingent negative variation: An electric sign of sensorimotor association and expectancy in the human brain. Nature. 1964;**203**(4943):380-384. DOI: 10.1038/203380a0

[3] Näätänen R, Gaillard AW, Mäntysalo S. Early selective-attention effect on evoked potential reinterpreted. Acta Psychologica. 1978;**42**(4):313-329

[4] Galvao-Carmona A, González-Rosa JJ, Hidalgo-Muñoz AR, Páramo D, Benítez ML, Izquierdo G, Vazquez-Marrufo M. Disentangling the attention network test: Behavioral, event related potentials, and neural source analyses. Frontiers in Human Neuroscience. 2014 Oct 13;**8**:813. DOI: 10.3389/fnhum.2014.00813

[5] Bostanov V, Keune PM, Kotchoubey B, Hautzinger M. Event-related brain potentials reflect increased concentration ability after mindfulness-based cognitive therapy for depression: A randomized clinical trial. Psychiatry Research. 2012 Oct 30;**199**(3):174-180. DOI: 10.1016/j.psychres.2012.05.031. Epub Jul 6, 2012

[6] Arjona A, Escudero M, Gómez CM. Cue validity probability influences neural processing of targets. Biological Psychology. 2016 Sep;**119**:171-183. DOI: 10.1016/j.biopsycho.2016.07.001

[7] Lytnev V, Fujiwara K, Kiyota N, Irei M, Toyama H, Yaguchi C. Postural control and contingent negative variation during transient floor translation while standing with the ankle fixed. Journal of Physiological Anthropology. 2016 Jul 25;**36**(1):7. DOI: 10.1186/s40101-016-0104-8

[8] Fan J, McCandliss BD, Sommer T, Raz A and Posner MI. Testing the efficiency and independence of attentional networks. Journal Cognitive Neuroscience. 2002;**14**:340-347. DOI: 10.1162/089892902317361886

[9] Callejas A, Lupiáñez J, Tudela P. The three attentional networks: On their independence and interactions. Brain and Cognition. 2004 Apr;**54**(3):225-227

[10] Fuentes LJ, Campoy G. The time course of alerting effect over orienting in the attention network test. Experimental Brain Research. 2008 Mar;**185**(4):667-672

[11] Kratz O, Studer P, Baack J, Malcherek S, Erbe K, Moll GH, Heinrich H. Differential effects of methylphenidate and atomoxetine on attentional processes in children with ADHD: An event-related potential study using the attention network test. Progress in Neuropsychopharmacology and Biological Psychiatry. 2012 Apr 27;**37**(1):81-89. DOI: 10.1016/j.pnpbp.2011.12.008. Epub Dec 29, 2011

[12] Tecce JJ. Contingent negative variation (CNV) and psychological processes in man. Psychological Bulletin. 1972;**77**(2):73-108. DOI: 10.1037/h0032177

[13] Frost BG, Neill RA, Fenelon B. The determinants of the non-motoric CNV in a complex, variable foreperiod, information processing paradigm. Biological Psychology. 1988;**27**(1):1-21. DOI: 10.1016/0301-0511(88)90002

[14] Segalowitz SJ, Davies PL. Charting the maturation of the frontal lobe: An electrophysiological strategy. Brain and Cognition. 2004 Jun;**55**(1):116-133

[15] Schmitt H, Ferdinand NK, Kray J. Age-differential effects on updating cue information: Evidence from event-related potentials. Cognitive, Affective, & Behavioral Neuroscience. 2014 Sep;**14**(3):1115-1131. DOI: 10.3758/s13415-014-0268-9

[16] Weerts TC, Lang PJ. The effects of eye fixation and stimulus and response location on the contingent negative variation (CNV). Biological Psychology. 1973;**1**(1):1-19. DOI: 10.1016/0301-0511(73)90010-0

[17] Giard MH, Perrin F, Pernier J & Bouchet P. Brain generators implicated in the processing of auditory stimulus deviance: A topographic event-related potential study. Psychophysiology. 1990;**27**:627-640. DOI: 10.1111/j.1469-8986.1990.tb03184.x

[18] Knott VJ, Lapierre YD, De Lugt D, Griffiths L, Bakish D, Browne M & Horn E. Preparatory brain potentials in major depressive disorder. Progress in Neuro-Psychopharmacology & Biological Psychiatry. 1991;**15**:257-262. DOI: 10.1016/0278-5846(91)90089-J

[19] Gaillard AW. Effects of warning-signal modality on the contingent negative variation (CNV). Biological Psychology. 1976 Jun;**4**(2):139-154

[20] Hultin L, Rossini P, Romani GL, Högstedt P, Tecchio F, Pizzella V. Neuromagnetic localization of the late component of the contingent negative variation. Electroencephalography and Clinical Neurophysiology. 1996;**98**:425-448

[21] Zappoli, R. Permanent or transitory effects on neurocognitive components of the CNV complex induced by brain dysfunctions, lesions, and ablations in humans. International Journal of Psychophysiology. 2003;**48**(2):189-220. DOI: 10.1016/S0167-8760(03)00054-0

[22] de Tommaso M, Difruscolo O, Sciruicchio V, Specchio N, Livrea P. Abnormalities of the contingent negative variation in Huntington's disease: Correlations with clinical features. Journal of Neurological Sciences. 2007 Mar 15;**254**(1-2):84-89. Epub Feb 14, 2007

[23] Zappoli R, Versari A, Arnetoli G, Paganini M, Muscas GC, Arneodo MG, Gangemi PF, Bartelli M. Topographic CNV activity mapping, presenile mild primary cognitive decline and Alzheimer-type dementia. Neurophysiologie Clinique. 1991 Dec;**21**(5-6):473-483

[24] van Deursen JA, Vuurman EF, Smits LL, Verhey FR, Riedel WJ. Response speed, contingent negative variation and P300 in Alzheimer's disease and MCI. Brain and Cognition. 2009 Apr;**69**(3):592-599. DOI: 10.1016/j.bandc.2008.12.007. Epub Jan 29, 2009

[25] Kuoppamäki M, Rothwell JC, Brown RG, Quinn N, Bhatia KP, Jahanshahi M. Parkinsonism following bilateral lesions of the globus pallidus: Performance on a variety of motor tasks shows similarities with Parkinson's disease. Journal of Neurology, Neurosurgery and Psychiatry. 2005 Apr;**76**(4):482-490

[26] Vázquez-Marrufo M, Galvao-Carmona A, González-Rosa JJ, Hidalgo-Muñoz AR, Borges M, Ruiz-Peña JL, Izquierdo G. Neural correlates of alerting and orienting impairment in multiple sclerosis patients. PLoS One. 2014 May 12;**9**(5):e97226. DOI: 10.1371/journal.pone.0097226. eCollection 2014

[27] Kirenskaya AV, Kamenskov MY, Myamlin VV, Novototsky-Vlasov VY, Tkachenko AA. The antisaccade task performance deficit and specific CNV abnormalities in patients with stereotyped paraphilia and schizophrenia. Journal of Forensic Science. 2013 Sep;**58**(5):1219-1226. DOI: 10.1111/1556-4029.12241. Epub Jul 30, 2013

[28] Oke S, Saatchi R, Allen E, Hudson NR, Jervis BW. The contingent negative variation in positive and negative types of schizophrenia. American Journal of Psychiatry. 1994 Mar;**151**(3):432-433

[29] Siniatchkin M, Gerber WD, Kropp P, Vein A. Contingent negative variation in patients with chronic daily headache. Cephalalgia. 1998 Oct;**18**(8):565-569; discussion 531

[30] Vazquez-Marrufo M, González-Rosa JJ, Vaquero E, Duque P, Escera C, Borges M, Izquierdo G, Gómez CM. Abnormal ERPs and high frequency bands power in multiple sclerosis. International Journal of Neuroscience. 2008 Jan;**118**(1):27-38

[31] Kane NM, Butler SR, Simpson T. Coma outcome prediction using event-related potentials: P(3) and mismatch negativity. Audiology & Neuro-Otology. 2000 May–Aug;**5**(3-4): 186-191

[32] Näätänen R, Paavilainen, P, Titinen H, Jiang D, Alho D. Attention and mismatch negativity. Psychophysiology. 1993;**30**(5):436-450

[33] Schröger E, Winkler I. Presentation rate and magnitude of stimulus deviance effects on human pre-attentive change detection. Neuroscience Letters. 1995 Jul 7;**193**(3):185-188

[34] Mullens D, Woodley J, Whitson L, Provost A, Heathcote A, Winkler I & Todd J. Altering the primacy bias – How does a prior task affects mismatch negativity? Psychophysiology. 2014;**51**:437-445

[35] Pazo-Alvarez P, Cadaveira F & Amenedo E. MMN in the visual modality: A review. Biological Psychology. 2003;**63**:199-236

[36] Kremláček J, Kreegipuu K, Tales A, Astikainen P, Põldver N, Näätänen R, Stefanics G. Visual mismatch negativity (vMMN): A review and meta-analysis of studies in psychiatric and neurological disorders. Cortex. 2016 Jul;**80**:76-112. DOI: 10.1016/j.cortex.2016.03.017

[37] Strömmer JM, Tarkka IM, Astikainen P. Somatosensory mismatch response in young and elderly adults. Frontiers in Aging Neuroscience. 2014 Oct 27;**6**:293. DOI: 10.3389/fnagi.2014.00293

[38] Cheour-Luhtanen M, Alho K, Kujala T, Sainio K, Reinikainen K, Renlund M, Aaltonen O, Eerola O, Näätänen R. Mismatch negativity indicates vowel discrimination in newborns. Hearing Research. 1995 Jan;**82**(1):53-58

[39] Cleary KM, Donkers FC, Evans AM, Belger A. Investigating developmental changes in sensory processing: Visual mismatch response in healthy children. Frontiers in Human Neuroscience. 2013 Dec 30;**7**:922. DOI: 10.3389/fnhum.2013.00922

[40] Cooper RJ, Todd J, McGill K and Michie PT. Auditory sensory memory and the aging brain: A mismatch negativity study. Neurobiology of Aging. 2006;**27**:752-762. DOI: 10.1016/j.neurobiolaging.2005.03.012

[41] Gaeta H, Friedman D, Ritter W, Cheng J. An event-related potential evaluation of involuntary attentional shifts in young and older adults. Psychology and Aging. 2001 Mar;**16**(1):55-68

[42] Alho K. Cerebral generators of mismatch negativity (MMN) and its magnetic counterpart (MMNm) elicited by sound changes. Ear and Hearing. 1995 Feb;**16**(1):38-51

[43] Lang HA, Nyrke T, Ek M, Aaltonen O, Raimo I, Naatanen R. Pitch discrimination performance and auditory event-related potentials. In: Brunia CHM, Gaillard AWK, Kok A, Mulder G, Verbaten MN, editors. Psychophysiological Brain Research. Vol. 1. Tilburg: Tilburg University Press; 1990; pp. 294-298

[44] Kraus N, McGee T, Carrell TD, Sharma A. Neurophysiologic bases of speech discrimination. Ear and Hearing. 1995 Feb;**16**(1):19-37

[45] Näätänen R, Sussman ES, Salisbury D, Shafer VL. Mismatch negativity (MMN) as an index of cognitive dysfunction. Brain Topography. 2014 Jul;**27**(4):451-466. DOI: 10.1007/s10548-014-0374-6

[46] Salisbury DF, Shenton ME, Griggs CB, Bonner-Jackson A, McCarley RW. Mismatch negativity in chronic schizophrenia and first-episode schizophrenia. Archives of General Psychiatry. 2002 Aug;**59**(8):686-694

[47] Oades RD, Wild-Wall N, Juran SA, Sachsse J, Oknina LB, Röpcke B. Auditory change detection in schizophrenia: Sources of activity, related neuropsychological function and symptoms in patients with a first episode in adolescence, and patients 14 years after an adolescent illness-onset. BMC Psychiatry. 2006 Feb 8;**6**:7

[48] Jung J, Morlet D, Mercier B, Confavreux C, Fischer C. Mismatch negativity (MMN) in multiple sclerosis: An event-related potentials study in 46 patients. Clinical Neurophysiology. 2006 Jan;**117**(1):85-93. Epub Dec 1, 2005

[49] Ilvonen TM, Kujala T, Kiesiläinen A, Salonen O, Kozou H, Pekkonen E, Roine RO, Kaste M, Näätänen R. Auditory discrimination after left-hemisphere stroke: A mismatch negativity follow-up study. Stroke. 2003 Jul;**34**(7):1746-1751. Epub Jun 19, 2003

[50] Särkämö T, Tervaniemi M, Laitinen S, Forsblom A, Soinila S, Mikkonen M, Autti T, Silvennoinen HM, Erkkilä J, Laine M, Peretz I, Hietanen M. Music listening enhances cognitive recovery and mood after middle cerebral artery stroke. Brain. 2008 Mar;**131**(Pt 3):866-876. DOI: 10.1093/brain/awn013

[51] Shafer VL, Morr ML, Datta H, Kurtzberg D, Schwartz RG. Neurophysiological indexes of speech processing deficits in children with specific language impairment. Journal Cognitive Neuroscience. 2005 Jul;**17**(7):1168-1180

[52] Gomot M, Giard MH, Adrien JL, Barthelemy C, Bruneau N. Hypersensitivity to acoustic change in children with autism: Electrophysiological evidence of left frontal cortex dysfunctioning. Psychophysiology. 2002 Sep;**39**(5):577-584

[53] Kuhl PK, Coffey-Corina S, Padden D, Dawson G. Links between social and linguistic processing of speech in preschool children with autism: Behavioral and electrophysiological measures. Developmental Science. 2005 Jan;**8**(1):F1-F12

[54] Vazquez-Marrufo M, González-Rosa JJ, Galvao-Carmona A, Hidalgo-Muñoz A, Borges M, Peña JL, Izquierdo G. Retest reliability of individual p3 topography assessed by high density electroencephalography. PLoS One. 2013 May 1;**8**(5):e62523. DOI: 10.1371/journal.pone.0062523

[55] Hruby T, Marsalek P. Event-related potentials--the P3 wave. Acta Neurobiologiae Experimentalis (Wars). 2003;**63**(1):55-63

[56] Sutton S, Tueting P, Zubin J. Information delivery and the sensory evoked potential. Science. 1967;**155**:1436-1439

[57] Polich J. P300 in clinical applications. In: Niedermayer E and Lopes de la Silva F, editors. Electroencephalography: Basic Principles, Clinical Applications and Related Fields. Baltimore-Munich: Urban and Schwartzenberger. 1999. pp. 1073-1091

[58] Katayama J, Polich J. Auditory and visual P300 topography from a 3 stimulus paradigm. Clinical Neurophysiology. 1999;**110**:463-468

[59] Polich J, Howard L, Starr A. Stimulus frequency and masking as determinants of P300 latency in event-related potentials from auditory stimuli. Biological Psychology. 1985 Dec;**21**(4):309-318

[60] Kutas M, McCarthy G, Donchin E. Augmenting mental chronometry: The P300 as a measure of stimulus evaluation time. Science. 1977;**197**:792-795

[61] Ohlrich ES, Barnet AB, Weiss IP, Shanks BL. Auditory evoked potential development in early childhood: A longitudinal study. Electroencephalography and Clinical Neurophysiology. 1978 Apr;**44**(4):411-423

[62] Goodin DS, Squires KC, Henderson BH, Starr A. Age-related variations in evoked potentials to auditory stimuli in normal human subjects. Electroencephalography and Clinical Neurophysiology. 1978 Apr;**44**(4):447-458

[63] O'Brien JH. P300 wave elicited by a stimulus-change paradigm in acutely prepared rats. Physiology & Behavior. 1982;**28**:711-713

[64] Harrison J, Buchwald J. Aging changes in the cat 'P300' mimic the human. Electroencephalography and Clinical Neurophysiology. 1985;**62**:227-234

[65] Paller, KA, Zola-Morgan, S, Squire, LR, Hillyard, SA. P3-like brain waves in normal monkeys and monkeys with medial temporal lesions. Journal of Behavioral Neuroscience. 1988;**102**:714-725

[66] Ehlers CL, Somes C. Long latency event-related potentials in mice: Effects of stimulus characteristics and strain. Brain Research. 2002 Dec 6;**957**(1):117-128

[67] Squires NK, Squires KC, Hillyard SA. Two varieties of long-latency positive waves evoked by unpredictable auditory stimuli in man. Electroencephalography and Clinical Neurophysiology. 1975 Apr;**38**(4):387-401

[68] Friedman D, Kazmerski V, Fabiani M. An overview of age-related changes in the scalp distribution of P3b. Electroencephalography and Clinical Neurophysiology. 1997 Nov;**104**(6):498-513

[69] Huang WJ, Chen WW, Zhang X. The neurophysiology of P300 – an integrated review. European Review for Medical and Pharmacological Sciences. 2015 Apr;**19**(8):1480-1488

[70] Desmedt JE. P300 in serial tasks: An essential post-decision closure mechanism. Progress in Brain Research. 1980;**54**:682-686

[71] Wickens C, Kramer A, Vanasse L, Donchin E. Performance of concurrent task: A psychophysiological analysis of the reciprocity of information-processing resources. Science. 1983;**221**:1080-1082

[72] Johnson VS, Miller DR, Burleson MH. Multiple P3s to emotional stimuli and their theoretical significance. Psychophysiology. 1986;**23**:684-694

[73] Donchin E, Coles MGH. Is the P300 manifestation of context updating? Behavioral and Brain Sciences. 1988;**11**:357-374

[74] Verleger R. Event-related potentials and cognition: A critique of the context updating hypothesis and an alternative interpretation of P3. Behavioral and Brain Sciences. 1988;**11**:343-427

[75] Kok A. Event-related potentials (ERP) reflections of mental resources: A review and synthesis. Biological Psychology. 1997;**45**:19-56

[76] Polich J. P300 clinical utility and control of variability. Journal of Clinical Neurophysiology. 1998;**15**:14-33

[77] Carretié L, Iglesias J, García T, Ballesteros M. N300, P300 and the emotional processing of visual stimuli. Electroencephalography and Clinical Neurophysiology. 1997 Aug;**103**(2):298-303

[78] Carillo-de-la Pena MT, Cadaveira F. The effect of motivational instructions on P300 amplitude. Neurophysiologie Clinique. 2000;**30**:232-239

[79] Vázquez-Marrufo M, González-Rosa J, Vaquero-Casares E, Duque P, Borges M, Izquierdo G. Cognitive evoked potentials in remitting-relapsing and benign forms of multiple sclerosis. Revista de Neurologia. 2009 May 1-15;**48**(9):453-458

[80] Vázquez-Marrufo M, Galvao-Carmona A, González-Rosa JJ, Hidalgo-Muñoz AR, Borges M, Ruiz-Peña JL, Izquierdo G. Neural correlates of alerting and orienting impairment in multiple sclerosis patients. PLoS One. 2014 May 12;**9**(5):e97226. DOI: 10.1371/journal.pone.0097226

[81] Comi G, Leocani L, Locatelli T, Medaglini S, Martinelli V. Electrophysiological investigations in multiple sclerosis dementia. Electroencephalography and Clinical Neurophysiology. Supplement. 1999;**50**:480-485

[82] Polich J, Ladish C, Bloom F. P3 assessment of early Alzheimer's disease. Electroencephalography and Clinical Neurophysiology. 1990;**77**:1

[83] O'Donnel BF, Squires NK, Martz MJ, Chen JR, Phay JR. Evoked potential changes and neuropsychological performance in Parkinson's disease. Biological Psychology. 1987;**24**:23-37

[84] Tanaka H, Koenig T, Pascual-Marqui RD, Hirata K, Kochi K, Lehmann D. Event-related potential and EEG measures in Parkinson's disease without and with dementia. Dementia and Geriatric Cognitive Disorders. 2000;**11**:39-45

[85] Daffner K, Mesulam M, Holcomb P, Calvo V, Acar D, Chabrerie A, Kikinis R, Jolesz F, Rentz D, Seinto L. Disruption of attention to novel events after frontal lobe injury in humans. Journal of Neurology, Neurosurgery and Psychiatry. 2000;**68**:18-24

[86] Alonso-Prieto E, Lvarez-Gonzalez Ma, Reyes-Verazain A. Use of event related potentials for the diagnosis and follow up of subclinical disorders of sustained attention in ischemic cerebrovascular disease. Revista de Neurologia. 2002;**34**:1017-1020

[87] Pritchard WS. Cognitive event-related potentials correlates of schizophrenia. Psychological Bulletin. 1986;**100**:43-66

[88] O'Donnel BF, McCarley R, Potts G, Salisbury D, Nestor P, Hirayasu Y. Identification of neural circuit underlying P300 abnormalities in schizophrenia. Psychophysiology. 1999;**36**:388-398

[89] Martín-Loeches M, Molina V, Munoz F, Hinojosa JA, Reig S, Desco M, Benito C, Sanz J, Gabiri A, Sarramea F, Santos A, Palomo T. P300 amplitude as a possible correlate of frontal degeneration in schizophrenia. Schizophrenia Research. 2001;**49**:121-128

Chapter 2

Application of P300 Event-Related Potential in Brain-Computer Interface

Ali Haider and Reza Fazel-Rezai

Additional information is available at the end of the chapter

http://dx.doi.org/10.5772/intechopen.69309

Abstract

The primary purpose of this chapter is to demonstrate one of the applications of P300 event-related potential (ERP), i.e., brain-computer interface (BCI). Researchers and students will find the chapter appealing with a preliminary description of P300 ERP. This chapter also appreciates the importance and advantages of noninvasive ERP technique. In noninvasive BCI, the P300 ERPs are extracted from brain electrical activities [electroencephalogram (EEG)] as a signature of the underlying electrophysiological mechanism of brain responses to the external or internal changes and events. As the chapter proceeds, topics are covered on more relevant scholarly works about challenges and new directions in P300 BCI. Along with these, articles with the references on the advancement of this technique will be presented to ensure that the scholarly reviews are accessible to people who are new to this field. To enhance fundamental understanding, stimulation as well as signal processing methods will be discussed from some novel works with a comparison of the associated results. This chapter will meet the need for a concise and practical description of basic, as well as advanced P300 ERP techniques, which is suitable for a broad range of researchers extending from today's novice to an experienced cognitive researcher.

Keywords: brain-computer interface, P300, electroencephalogram, event-related potential, paradigm and human factors

1. Introduction

Human brain is the most complex organ of the body and it is at the center of the driving block of human nervous system. In fact, more than 100 billion nerve cells are interconnected to build the functionality of human brain. Such a complicated architecture allows the brain to control the body as well as carry out the executive functions, such as making reasons,

INTECH
open science | open minds

© 2017 The Author(s). Licensee InTech. This chapter is distributed under the terms of the Creative Commons Attribution License (http://creativecommons.org/licenses/by/3.0), which permits unrestricted use, distribution, and reproduction in any medium, provided the original work is properly cited.

processing thoughts, and planning for next tasks. Interestingly, electrophysiology and hemodynamic response are the two techniques that have been used to study this complex organ to understand the mechanism the brain applies to finish works. Typically, electrophysiological measurements are performed by placing electrodes or sensors on the biological tissue [1, 2]. In neuroscience and neuro-engineering, the electrophysiological techniques are used for studying electrical properties by measuring the electrical activities of neurons in the form of electroencephalogram (EEG). EEG may be measured by two different approaches: invasive and noninvasive. Invasive procedures need a surgery to place the EEG sensor deep under the scalp. In comparison, noninvasive procedure places the electrodes on the scalp. One of the ways to study the brain is to stimulate it by presenting a paradigm.

The event-related potential (ERP) was first reported by Sutton [3]. An ERP is an electrophysiological response or electrocortical potentials triggered by a stimulation and firing of neurons. A specific psychological event or a sensor can be employed to generate the stimulation. In general, visual, auditory, and tactile are three major sources of ERP stimulation. For instance, ERP can be elicited by a surprise appearance of a character on a visual screen, or a "novel" tone presented over earphones, or by sudden pressing of a button by the subject, including myriad of other events. Presented stimulus generates a detectable but time-delayed electrical wave in EEG. EEG is recorded starting from the time of presenting the stimulus to the time when EEG settles down. Depending on the necessity, simple detection method such as ensemble averaging or advanced processes such as linear discriminant analysis or support vector machine algorithms are applied on EEG to measure the ERP. This chapter discusses the application of ERP in brain-computer interface (BCI) where P300 wave is of particular interest. ERP is time-locked to an event and appears as a series of positive and negative voltage fluctuation in the EEG that is referred to as P300 components.

2. P300 waveform

P300 is a form of visually evoked potential (VEP) and P300 ERP is embedded within the EEG signal recordable from the scalp of human brain. Depending on the components appearance following the eliciting event, the P300 can be divided into exogenous and endogenous. Early (exogenous) components are distributed over first 150 ms, whereas longer latency (endogenous) components elicit after 150 ms. Although the P300 positive deflection occurs in the EEG about 300 ms after an eliciting stimulus is delivered (which is the major reason it is termed as P300), latency can be within the range from 250 to 750 ms.

Although the actual origin of the P300 is still unclear, it is suggested that P300 is elicited by the decision making or learning that a rare event has occurred, and some things appear to be learned if and only if they are surprising [4]. The variable latency is associated with the difficulty of the decision making. In addition, the largest P300 responses are obtained over parietal zone of human head while it is attenuated with the electrodes that are gradually placed farther from this area.

To generate the P300 ERP, three different types of paradigms are being used: (1) single-stimulus, (2) oddball, and (3) three-stimulus paradigm. In each case, the subject is instructed to follow the occurrence of the target by pressing a button or mentally counting [5]. **Figure 1** presents these paradigms [5, 6]. The single-stimulus paradigm irregularly presents just one type of stimuli or target with zero occurrence of any other type of target. A typical oddball paradigm can be presented to the subject with a computer screen, a group of light-emitting diodes (LEDs), or other medium to generate a sequence of events that can be categorized into two classes: frequently presented standard (nontarget or irrelevant) and rarely presented target stimuli [7]. In an oddball paradigm, two events are presented with different probabilities in a random order, but only the irregular and rare event (the oddball event) embosses the P300 peak into the EEG about 300 ms after the stimulus onset. The three-stimulus paradigm is a modified oddball task which includes nontarget distractor (infrequent nontarget) stimuli in addition to target and standard stimuli. The distractor elicits P3a which is large over the frontal/central area [8]. In contrast, target elicits a P3b (P300), which is maximum over the parietal electrode sites. Though P3a and P3b are subcomponents of P300, P3a is dominant in the frontal/central lobe with a shorter latency and habituates faster [9].

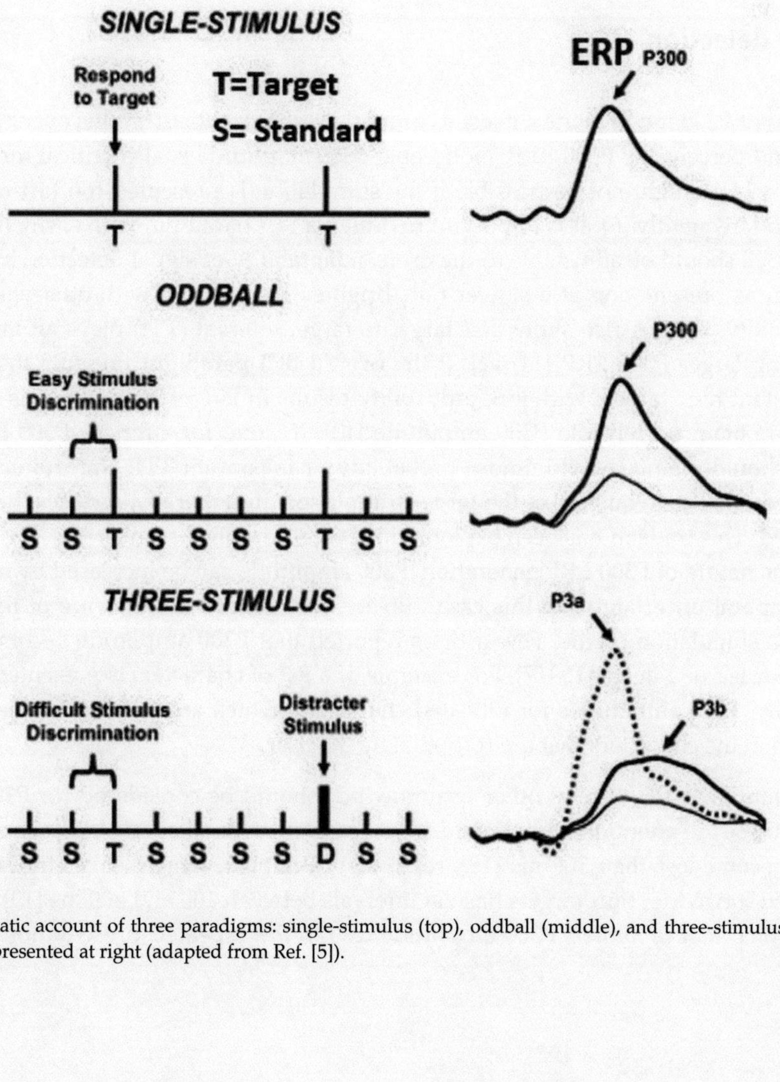

Figure 1. Schematic account of three paradigms: single-stimulus (top), oddball (middle), and three-stimulus (bottom). Elicited ERP is presented at right (adapted from Ref. [5]).

3. P300 BCI

Humans' ability to communicate to each other plays a critical role in building relationship with society and others. With the advent of modern logistics and necessities, communication between people has become richer and more complex than any other time of human history. Furthermore, as brain science and computer technologies mature, it is critical to have an ultimate interaction interface that will develop a direct communication between the user's brain and a computer; in other words, a BCI system which facilitates to build a real-time communication between a user and a computer system. The core purpose of a BCI is to detect brain activity in EEG and communicate that activity to a computer or electronic device. BCI allows a user to voluntarily send messages or control commands bypassing the brain's natural output pathways. There have been different approaches for BCI. P300 BCI is a safe and noninvasive system, which requires the user to wear a small head cap carrying a set of electrical probes to detect brain P300 ERP. The P300 BCI has many potential advantages over many other input modes [10].

4. P300 detection

Detection of P300 requires the subject to properly recognize the stimulus event to generate a strong and perceivable P300 ERP. Noticeable P300 amplitude is also critical for information transfer, which might not be possible if the stimulation is presented too fast or the targets appear too frequently [6]. It is important to design a BCI paradigm with easily discriminable stimuli. BCI should be adjustable to the users' adaptability of signal detection by controlling the stimulus presentation at a slower rate, brighter intensity, or with otherwise increasing perceptibility. Studies also show that target-to-target interval (TTI) plays an important role in evoking larger P300 ERP [11, 12]. If the overall BCI paradigm presents the stimulation at a constant rate, targets with low probability results in longer TTL, which is also a useful means to obtain perceivable P300 amplitude [13]. In sum, for stronger P300 ERP, the BCI system should maintain a minimum probability or maximum TTI. Unfortunately, such an action reduces the frequency of the target stimulation and, thereby, reduces the overall system speed. This tradeoff has been explored in several early BCI studies [14]. It is evident that due to the nature of P300 ERP generation, P300 amplitude can be increased by incorporating high temporal uncertainty. In this case, subjects are completely unaware of the exact time when the stimulation occurs. Few articles reported that P300 amplitude becomes larger for familiar or learned items [15–17]. For example, if a list of characters is presented to a subject repeatedly, P300 amplitudes for repeated characters (which are recalled by the subject) are higher than the characters that are forgotten by the user.

In addition, there are several other factors which should be considered for P300 detection. Among these are attentional blink, which occurs in case the intervals between two different targets become less than 500 ms [18]; repetition blindness, which leaves the second target unnoticed if two identical targets flash at intervals between 100 and 500 ms [19]; and habituation, which makes fainted P300 amplitude due to the repeated presentation of the same stimulus [20]. Apart from this, human factors such as motivation, fatigue, and user comfortability affect the performance and accuracy of the P300 BCI [21–23], which should be considered in the design of paradigms.

5. Signal processing methods

A P300-based BCI measures EEG signals from the human scalp and processes them in real time to detect P300 ERP that reflect the subject's intent. As noted earlier, P300-evoked potential is elicited as positive EEG peaks in reaction to infrequent or irregular appearance of stimuli. As the EEG signals are typically on the order of 100 microvolts, appropriate signal processing strategy is critical in revealing the electrical information and relevant complex issues in relation to the distinctive cognitive functions. Moreover, optimization of accuracy in P300 detection and enhancement of the system speed heavily depends on a suitable signal processing scheme.

EEG-based BCI system can have three stages to process signals: preprocessing, feature extraction, and detection and classification of P300. Preprocessing is accomplished after data acquisition but before extracting any feature. Preprocessing is an important step which leaves the significant information intact while amplifying EEG signals and simplifying subsequent processing operations. It is also important to note that the classifier performance depends greatly on an efficient data preprocessing stage [24]. Signal strengthening ensures signal quality by improving the so-called signal-to-noise ratio (SNR). Presence of background noise may bury the interesting brain patterns into the rest of signal making it difficult to detect P300 response resulting in a bad or small SNR. On the other hand, P300 detection and classification becomes easier when the input EEG signal has high SNR. After acquiring the EEG signal from microelectrodes or macroelectrodes, the electrical information is amplified by a factor of as high as 5000–10,000 and converted from an analog to a digital signal. Though analog to digital A/D conversion can be done at a rate of few GHz, human brain does not operate that fast to justify such a high sampling frequency. EEG data is typically sampled at 256 Hz which satisfies the Nyquist sampling theorem as this rate is larger than two times the maximum frequency generated by cognitive actions, yet low enough to avoid irrelevant data [25]. To realize the high SNR, bandpass filtering is utilized to remove the DC bias and high frequency noise. Sometime, researchers also combine transformation and filtering techniques and apply to remove or abate signal components that are not of interest for the application [26, 27]. As AC current is usually of 50–60 Hz, depending on the particular living zone of the globe, a notch filter at either 50–60 Hz is used to remove power line effect on EEG. During filter set up, it should be kept in mind that certain types of artifact occur at known frequencies and cognitive activity usually limits itself in the 3–40 Hz range.

Once the EEG is preprocessed, variety of approaches can be applied to extract the features and classify the P300 ERP. A calibration session is exploited to develop these feature vectors. Before classification test and actual use of the P300 BCI, the classifier is trained and supervised using a classification algorithm and the feature vectors labeled as "target" and "nontarget" [27]. On the other hand, during the classification task the feature vectors corresponding to known stimuli are submitted to a trained classifier. The trained classifier discriminates the brain response best resembling to a target stimulus from nontarget stimulus. In case of a P300 Speller, the classifier detects the letter with a maximum probability [10, 28–30].

Different methods have been employed for feature extraction such as discrete wavelet transform [31], independent component analysis [32, 33], and principal component analysis [34]. As stated earlier, extracted features are given as input to the EEG classifiers for P300 ERP identification and classification applying different classification methods. Linear discriminant analysis (LDA) is a popular pattern classification technique used by Guger et al. [35]. Stepwise linear discriminant analysis (SWDA) has evolved from LDA classification method which uses only selective features. Farwell and Donchin used SWDA to classify the ERP using individual averages for rows and columns of a 6 × 6 row/column paradigm [25]. Some classification methods apply machine learning technique for the P300 detection such as support vector machine (SVM) [36]. SVM takes advantage of small data size to give high throughput at high transfer rate. However, LDA outperforms SVM classifiers for the P300 detection if the input data is comparatively larger in size [37]. Moreover, many BCI groups have exercised their study with other classifiers such as Bayesian linear discriminant analysis (BLDA), Pearson's correlation method (PCM), linear support vector machine (LSVM), and Gaussian support vector machine (GSVM) [24, 38, 39]. Although different features and classifiers have been compared, there has not been a comprehensive comparison of all different features extraction

Methods	System performance
Discrete wavelet transform (DWT); 6 × 6 targets on the menu; 36 feature vectors; feature vectors were continually ranked and either a correlation/threshold was used to select a cell	7.8 characters/min and 80% accuracy; Accuracy >90% for 5 subjects [31]. 2.3 characters/min [25]
Genetic algorithm (GA); high resource consumption; possible premature convergence	Variable accuracy, 34~90% [40]
Bayesian analysis, Bayesian linear discriminant analysis (BLDA); feature vector is labeled to the class to which it has the highest probability	Transfer rate of 7 commands/min with 95% false positive classification accuracy [41]
Linear discriminant analysis (LDA); simple, low computation	Accuracy for the able-bodied subjects was on average close to 100% and the best classification accuracy for disabled subjects was on average 100%. 15.9 bits/min for the disabled subjects and 29.3 bits/min for the able-bodied subjects. Accuracy varies with electrodes 4–32 [42]
Support vector machine (SVM); linear and nonlinear (Gaussian) modalities, faster processing	96.5% accuracy [43]. Accuracies are 66, 69, and 72% for LDA, neural networks, and SVM, respectively [44] Accuracy 84.5% and information transfer rate of up to 84.7 bits/min [45]
Maximum likelihood (ML); feature detection using a priori knowledge, uses thresholds for a set of classes	Accuracy 90% with a communication rate of 4.19 symbols/min [33]

Table 1. Summary of the signal processing methods.

and classification methods applied to the same data set. However, a research group in Ref. [24] examined multiple feature extraction and classification methods applying to the same data set. This study found that SWDA and Fisher's linear discriminant (FLD) yield the best overall classification performance in comparison to any other classifier. Most frequently used signal processing methods have been described in **Table 1** with reference to the relevant study.

6. Advantages of P300 BCI

There are properties of P300 BCI that make it attractive in many applications including daily life usages [35]: (1) the typical P300 BCI can be controlled with high accuracy; (2) the P300 BCI classifier offers fast response; (3) it may be used in gaming applications where an even shorter calibration can be used if classifier accuracy is not critical; (4) almost all healthy people and many severely paralyzed patients are able to use the P300 BCI [20, 46, 47]; (5) unlike other BCI (e.g., the motor imagery-based BCI), no special training is needed to operate this BCI; and (6) P300 BCI is noninvasive, calibration time is limited to few minutes, and it is effective for most users and more than 90% users feel comfortable with this system.

7. Advancement of P300 BCI

In 1988, Farwell and Donchin furnished a seminal study to demonstrate the potential of P300-based communication with a P300 ERP-based speller [25]. Since then, many P300-based BCI systems have followed this as a benchmark for P300 ERP application. However, until the year 2000, it drew very little attention from research community and no P300 BCI peer-reviewed papers were available before this time [31]. Later on, some researchers have explored the offline analysis of previously collected BCI data resulting in a moderate increase in P300 BCI articles [32, 48]. However, in the last decade, extensive studies and interests have been observed with a particular focus on the importance of P300-based BCIs. In addition to improving information transfer rate, current P300-based BCI mainly explores new electrode montage, paradigms, and applications to increase the performance of the BCI systems which can also assist disabled users in home settings [10, 28, 47]. For example, effect of contrast and color, modified stimulus presentation, enhanced users' attention, and new paradigms [49, 50] for eliciting the P300 have been introduced as new ways to improve ERPs and its classification [51, 52]. As a result of strong interest in P300 BCI, BCI community has experienced a high volume of research works involving P300 ERP in past 6 years. **Figure 2** portrays the number of peer-reviewed journal publications that were accounted by PubMed and Scopus search engines from 2000 to 2016 with the phrase "[(BCI OR Brain Computer Interface) AND (P300 OR P3)]." Even though conference proceedings were not included in the result, novel studies were reflected by the large number of articles. Although, in particular, BCIs are still slower than normal electronic input devices, such as the mouse or game controllers, endless exploration of new options with considerable success promises a P300 BCI system with substantial increase in BCI speed and accuracy and, thereby, extending P300 BCIs to new applications.

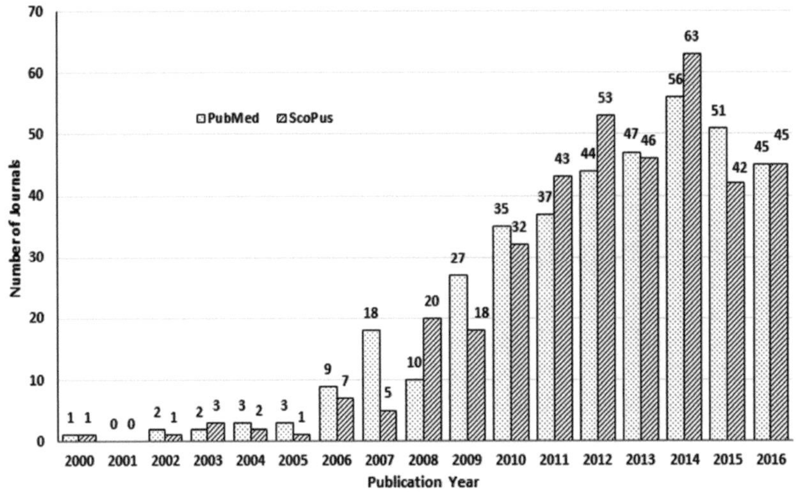

Figure 2. Number of published journal papers in PubMed and Scopus from 2000 to 2016 when the searching keyword "[(BCI OR Brain Computer Interface) AND (P300 OR P3)]" was used.

8. Applications of P300 systems

A P300 BCI is particularly suitable for selection applications [53]. For instance, the most typical application of P300 BCI is P300 speller. In such an arrangement, the visual paradigm is made up of a matrix consisting of letters of the alphabet. Depending on the requirement, a speller can be optimized for quick selection or accuracy of the spelled letters. Similarly, other P300 BCI investigations have made extensive progress to develop other attractive applications such as painting artwork, controlling smart home, designing games, stroke rehabilitation, lie detection, and furnishing Internet tasks [54]. However, recognizing the importance of P300 speller, a detail description of P300 speller is presented in the following sections.

8.1. Smart home

A smart home populates different electronic devices which can be controlled using a P300 BCI. A virtual reality-based smart home was the test-bed of such BCI application [35]. This BCI system allowed to execute a group of modest controlling commands such as moving the cart or wheelchair, receiving or making phone calls, operating television, switching the light on and off, playing song in multimedia player, or controlling the doors and windows [35].

8.2. Internet use

P300 BCI can be used to select the Internet keys to provide assistance to amyotrophic lateral sclerosis (ALS) patients browsing the websites. Subjects can surf through Internet pages and select the desired links to browse the Internet or read the news [55].

8.3. Painting task

It was observed by the researchers that performing natural tasks bring better quality to life in ALS patients. A P300 BCI application known as "Brain Painting" (BP) offers a medium of entertainment for the patients by improving their playful mood [56].

8.4. BCI gaming

P300 BCI has been used to design paradigm to control simple games that do not require strong time constraints such as to play chess [57]. Other popular games are MindGame [58], Bacteria Hunt [59], Brain Invaders [60], etc. In MindGame, the users move depends on the brain response; if P300 ERP is stronger, the game character can move larger distance. In Bacteria Hunt, users can change the color of the image, or enlarge, or rotate it. Similarly, in Brain Invaders, the user needs to select appropriate target arms to destroy the aliens, which make the game interesting to the video gamers. As no training is required to start playing simple P300 BCI games as mentioned here, it can be useful to familiarize individuals to the BCI tools. In fact, proper design to utilize the P300 wave's strong dependence on attention would allow the scientists to study attention training and effects of engaging in a particular task.

8.5. Stroke rehabilitation

One of the sufferings of poststroke patients is that they would like to say what they want but trouble of cortical circuits will not allow them to express it through natural motor pathways. P300 BCI paradigm was used to provide a communication channel to the participants diagnosed with poststroke aphasia. P300 BCI not only allowed to activate their language circuits, but also made their poststroke recovery faster [23].

8.6. Lie detection

Different brain regions work together and generate activities to process deceptive information which elicits P300 ERP in the brain signal. The concealed information can be identified through the concealed information test (CIT) [61]. Most of the earlier experiments with lie detectors used just a few channels limiting the number of EEG features to classify these two types of information [61–64]. These studies mostly used an oddball paradigm using three different types of stimuli: target, probe, and irrelevant. Like a typical P300-based system, the targets are presented rarely though they are usually made of irrelevant items which are presented in the paradigm to ensure participants' cooperation in discriminating the target items from others. On the contrary, the irrelevant items are presented frequently, but they are neither related to the criminal act nor related to the experimental task. The underlying principle of the item is that subjects will have different responses to stimuli according to their crime-relevant status. The probes are the critical detail stimuli under investigation which appear infrequently. Probes elicit P300 only for subjects who are knowledgeable or deceiving the information. Otherwise they act similarly as irrelevant for the subject. However, to ensure reliable differences between liars and truth-tellers it is important to engage multiple channels resulting in ERP features from different brain areas. One study investigated the functional connectivity of the brain network under deception condition [65]. They found the correlation between different EEG signals from multiple channels to understand the interactions between the brain regions and functional connectivity. Their results suggest that incorporation of additional features helps separating innocent group from the liars with about 90% accuracy.

8.7. P300 speller

Perhaps most important and popular use of P300 BCI is P300 speller. BCI speller has been utilized as a communication tool for the last two decades by people suffering from various neuromuscular disorders such as ALS, brainstem stroke, brain or spinal cord injury, cerebral palsy, muscular dystrophies, multiple sclerosis, and other impaired patients who are unable to use the normal neuronal pathway [66]. Persistent research in BCI to improve the accuracy and speed of P300 speller has resulted in numerous P300 stimuli presentation paradigms. They are discussed in details in the following sections.

8.8. Row/column (RC) paradigm

The Farwell and Donchin matrix speller paradigm was the first BCI row-column speller (**Figure 3**). They used an alphabetical square matrix interface to produce P300 in EEG [25]. Rows and columns of this 6 × 6 matrix were constructed with alphanumeric characters. These characters are flashed randomly following either a row or a column and the subject is asked to mentally count the number of times that the attended character is flashed. During the brain signal measurement in the parietal area, the P300 ERP appears in EEG as evoked response. However, the nonflashing rows and columns do not generate P300. Due to the nature of the stimulation mechanism and to increase the accuracy of detection, the P300 system requires multiple trials to reach an acceptable accuracy. In practice, the nontarget rows also generate P300 for a very short amount of time but the amplitude is too faint to detect. The computational device can determine the target row and column after averaging several P300 ERP responses. Due to the averaging task, it may take a longer time to detect a character.

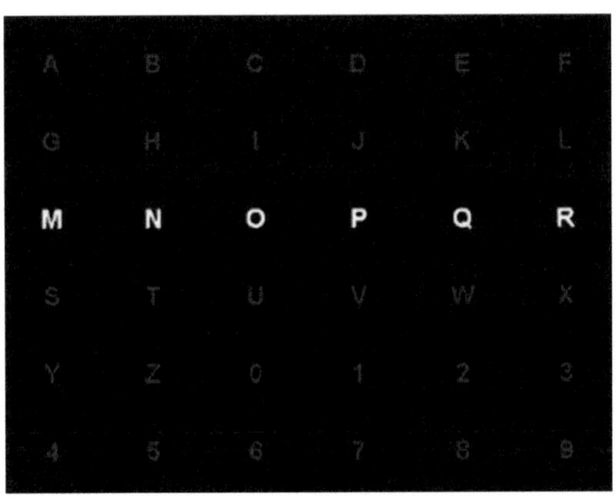

Figure 3. The row-column (RC) paradigm. One row (MNOPQR) is flashing.

In general, reducing the number of characters would eliminate the longer detection time but not without a loss in spelling characters option. So far, this is the mostly used and discussed P300 speller in BCI community.

8.9. Single character (SC) paradigm

This is possibly one of the simplest spellers designed so far. It randomly flashes one character at a time with very short interflash interval (**Figure 4**). This paradigm also uses a 6 × 6 alphanumeric matrix like the RC paradigm. It was reported that the RC paradigm takes less time than the SC paradigm to flash all the characters at least once. Nevertheless, in Ref. [35] it was noticed that if the number of flashes is constant, the SC speller produces stronger P300 ERP than the RC speller.

8.10. Region-based (RB) paradigm

The major idea behind the region-based (RB) paradigm is to distribute the characters in larger area than the RC paradigm. Here, choice of an object is split in dual selection levels which decreased the near-target effect and human error and adjacency problem significantly [49, 67]. In this paradigm, space of the visual paradigm is divided into seven different regions (**Figure 5**). The desired characters are split into seven groups and each group is placed into a single region as shown in **Figure 5**. For any given spelling task, user has few seconds to focus on the characters before the action of each level. This action produces the P300 ERP for the first-level target. It is important to note that, instead of the rows and columns as in Farwell and Donchin RC paradigm, regions are flashed in a random order by repeatedly changing its color between black and white. Choice of color needs to be justified with the purpose of the application. For simple spelling task, common black-white transition usually outcomes better contrast. Combined action of these two levels is needed to detect a single character. For instance, the first level is used to select the desired region containing the character of interest while the second level is used to select the target character from the enlarged region. Expansion at the second level largely increases the intercharacter space so that each character

Figure 4. Single character (SC) paradigm: single character (M) is flashed.

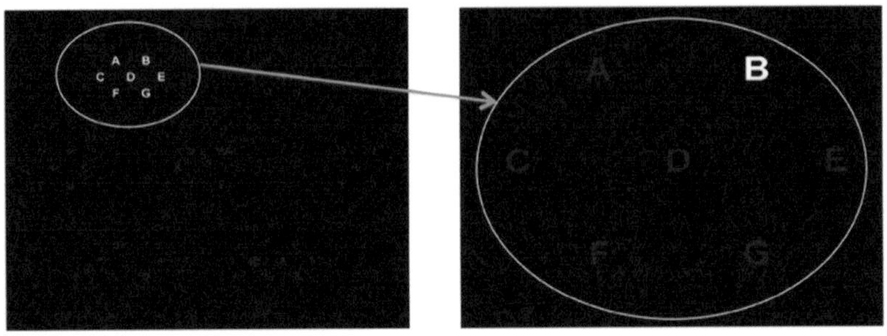

Figure 5. Region-based (RB) paradigm with the locations of seven regions, where a region of a set of seven characters "ABCDEFG" in level 1 is expanded in level 2 for spelling a single character "B".

is highly visible to the user. Each time a target is flashed, a strong P300 ERP potential is expected in the EEG wave. Although Farwell and Donchin paradigm allows one to spell 36 characters, the use of seven-region RB paradigm allows to spell 49 characters. In addition, this arrangement allows to manipulate and distribute the characters spatially on the screen considering their probability of linguistics use in a word. As paralyzed people need to spell the desired word with a minimum movement, the arrangement of the letters can be adjusted accordingly to optimize the performance [10]. Later on, RB paradigm was modified by implementing the findings about the probability of characters' usage [68]. In fact, it was developed considering the frequency of use of the characters. The list of characters used in seven regions in the first level of this modified edition is presented in **Table 2**. In a comparative study, it was found that the overall spelling accuracies averaged for the same set of subjects, trials, and characters for RC, SC, and two variations of RB paradigms were 85, 72.2, 86.1, and 90.6%, respectively [14, 69]. It is interesting to note that other than P300 BCI, application of RB paradigm has been extended to other BCI modalities too [54, 70–72].

8.11. Checkerboard (CB) paradigm

A standard RC presentation method has couple of limitations which were addressed by the CB paradigm (**Figure 6**). First of all, in checkerboard paradigm in Ref. [50], row-column

Region	First-level characters
Region 1	E T A O N R I
Region 2	S H D L F C M
Region 3	U GY P W B V
Region 4	K X J Q Z 1 2
Region 5	3 4 5 6 7 8 9
Region 6	0 / * − + . ?
Region 7	" ! @ # $ % &

Table 2. List of characters in each region in the first-level of region-based paradigm.

http://dx.doi.org/10.5772/intechopen.69309

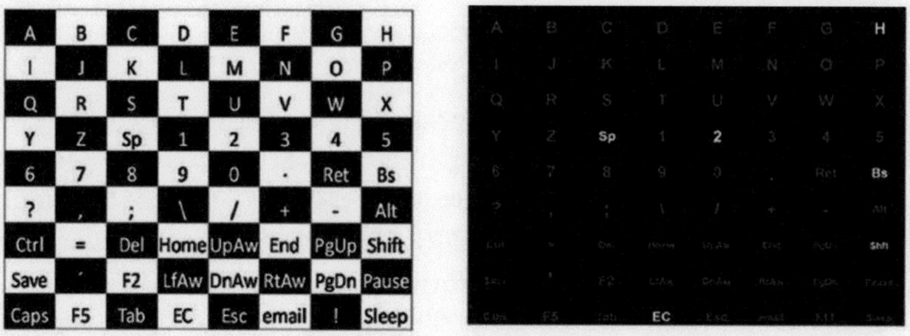

Figure 6. Standard checkerboard (CB) paradigm with a 9 × 8 matrix (adapted from Ref. [50]).

paradigm was modified to eliminate the error caused by adjacency problem as it was discussed by Fazel-Rezai [73] as human error in P300 BCI. So the error resulting from the nontarget items receiving apparent target responses is reduced to a great extent. Second, 72 characters in a standard 9 × 8 matrix are distributed in two virtual levels where each level is a virtual checkerboard, a modification to row/column (RC) paradigm implemented in a checkerboard (CB) paradigm. This eliminates the chances of the same character flashing twice in succession, thereby increasing the time between successive flashes of a target character.

9. Future use of P300 BCI

There exist many future directions to improve the information throughput in P300 BCIs, which also equally true for many other types of BCI systems. To uncover the applications of P300 ERP to other modalities, underlying physiological mechanism and brain response in each of the particular application need to be carefully investigated. For example, study to unfold more insight of the cognitive process showed that neurofeedback can be applied to augment the cognitive diagnosis [74]. In order to increase the BCI accuracy, error correction mechanisms can be incorporated into the BCI system. It will also increase the user acceptability of P300 BCI. Although improving information throughput of BCI is of paramount importance, many other aspects of BCIs also demands substantial consideration. For example, future BCIs need to be faster, inexpensive, and easy to use. Fortunately, BCI community comprises of many other disciplines, such as engineering, cognitive and neuroscience, semantics, mathematics, psychology, clinical science, and software writing. Eventually, scientists and researchers from various avenues continuously help finding a universal platform for BCI development utilizing available resources free for academic research. In particular, future expansion of BCI application depends a lot on the thorough investigation of users' comfort in using BCI. So different conditions should be well explored to find reasons behind why most users may or may not like a BCI system or paradigm. Many articles have introduced questionnaires and surveys to learn the comfort zone of the P300 BCI users [52, 56]. To promote the use of BCI to the target users with new applications, record and study of the human factors should be employed.

10. Conclusion

An ERP is a change in voltage which is time-locked to a specific sensory, motor, or cognitive event. ERP provides a distinctive pattern as an indication of how the stimulus is processed. Many BCI applications have been developed based on ERP as a response to stimulus. Among these, P300-based BCI is the most prominent ERP BCI. Over the last two decades, countless P300 BCI works have exploded beyond laboratory experiments with the help of modern high-speed computational and sensor technologies. Because of its noninvasive nature and stable performance, P300 applications range from the potential improvement of the lifestyle to the financial benefits. In fact, fundamental research on recording hardware, signal processing methods, stimulus presentation parameters, supporting interaction paradigm, and neurophysiology will further refine the P300-based BCI design. Though a BCI design is accomplished with keeping a specific application in mind, further insightful study and research can revive opportunities toward exploring other usability areas which are still not unearthed. This chapter has covered several aspects and applications of P300 ERP in BCI research. The interfacing paradigm of a P300 BCI can be designed to capture the ERP-evoked potentials in a manner so that many human factors are properly taken care of to diminish their overall negative impact. Many new applications are also emerging with efficient design of the control interface and associated signal processing scheme.

Author details

Ali Haider and Reza Fazel-Rezai*

*Address all correspondence to: reza.fazelrezai@engr.und.edu

Department of Electrical Engineering, Biomedical Engineering Research Complex, University of North Dakota, Grand Forks, USA

References

[1] Scanziani M, Häusser M. Electrophysiology in the age of light. Nature. 2009;**461**(7266): 930-939

[2] Shibasaki H. Human brain mapping: Hemodynamic response and electrophysiology. Journal of Clinical Neurophysiology. 2008;**119**(4):731-743

[3] Sutton S, Tueting P, Zubin J, John ER. Information delivery and the sensory evoked potential. Science (80-). 1967;**155**(3768):1436-1439

[4] Donchin E. Surprise!...Surprise? Psychophysiology. 1981;**18**:493-513

[5] Johnson R. The amplitude of the P300 component of the event-related potential: Review and synthesis. Advanced Psychophysiology. 1988;**3**:69-137

[6] Johnson R. A triarchic model of P300 amplitude. Psychophysiology. 1986;**23**:367-384

[7] Donchin E, Coles MGH. Is the P300 component a manifestation of context updating? Behavioral And Brain Sciences. 1988;**11**(3):355-425

[8] Comerchero MD, Polich J. P3a and P3b from typical auditory and visual stimuli. Clinical Neurophysiology. 1999;**110**(1):24-30

[9] Knight R. Contribution of human hippocampal region to novelty detection. Nature. 1996;**383**(6597):256-259

[10] Fazel-Rezai R, Allison BZ, Guger C, Sellers EW, Kleih SC, Kübler A. P300 brain computer interface: Current challenges and emerging trends. Frontiers in Neuroengineering. 2012;**5**(July):14

[11] Polich J. Task difficulty, probability, and inter-stimulus interval as determinants of P300 from auditory stimuli. Electroencephalography and Clinical Neurophysiology Evoked Potentials. 1987;**68**(4):311-320

[12] Duncan-Johnson CC, Donchin E. The P300 component of the event-related brain potential as an index of information processing. Biological Psychology. 1982;**14**(1-2):1-52

[13] Gonsalvez CL, Polich J. P300 amplitude is determined by target-to-target interval. Psychophysiology. 2002;**39**(3):388-396

[14] Fazel-Rezai R, Gavett S, Ahmad W, Rabbi A, Schneider E. A comparison among several P300 brain-computer interface speller paradigms. Clinical EEG and Neuroscience. 2011;**42**(4):209-213

[15] Noldy N, Stelmack R, Campbell K. Event-related potentials and recognition memory for pictures and words: The effects of intentional and incidental learning.Psychophysiology. 1990;**27**:417-428

[16] Fabiani M, Karis D, Donchin E. Effects of mnemonic strategy manipulation in a Von Restorff paradigm. Electroencephalography and Clinical Neurophysiology. 1990;**75**(2): 22-35

[17] Fabiani M, Karis D, Donchin E. P300 and recall in an incidental memory paradigm. Psychophysiology. 1986;**23**(3):298-308

[18] Raymond JE, Shapiro KL, Arnell KM. Temporary suppression of visual processing in an RSVP task: An attentional blink? Journal of Experimental Psychology: Human Perception and Performance. 1992;**18**:849-860

[19] Kanwisher NG. Repetition blindness: Type recognition without token individuation. Cognition. 1987;**27**(2):117-143

[20] Sellers EW, Donchin E. A P300-based brain-computer interface: Initial tests by ALS patients. Clinical Neurophysiology. 2006;**117**(3):538-548

[21] Gavett S, Wygant Z, Amiri S, Fazel-Rezai R. Reducing human error in P300 speller paradigm for brain-computer interface. In: Proceedings of the Annual International

Conference of the IEEE Engineering in Medicine and Biology Society (EMBS), San Diego, CA. 2012. p. 2869-72. DOI: 10.1109/EMBC.2012.6346562

[22] Rabbi AF, Zony A, de Leon P, Fazel-Rezai R. Mental workload and task engagement evaluation based on changes in electroencephalogram. Biomedical Engineering Letters. 2012;**2**(3):139-146

[23] Kleih SC, Gottschalt L, Teichlein E, Weilbach FX. Toward a P300 based brain-computer interface for aphasia rehabilitation after stroke: Presentation of theoretical considerations and a pilot feasibility study. Frontiers in Human Neuroscience. 2016;**10**:547

[24] Krusienski DJD, Sellers EEW, Cabestaing F, Bayoudh S, McFarland DJ, Vaughan TM, et al. A comparison of classification techniques for the P300 Speller. Journal of Neural Engineering. 2006;**3**(4):299-305

[25] Farwell LA, Donchin E. Talking off the top of your head. Electroencephalography and Clinical Neurophysiology. 1988;**70**(6):510-523

[26] Azinfar L, Ravanfar M, Kim E, Amiri S, Fazel-Rezai R. EEG Channel Optimization Based on Differential Evolutionary Algorithm for BCI Application. In: Fifth International Brain-Computer Interface Meeting. Pacific Gove, CA.; 2014. Graz University of Technology Publishing House; DOI: 10.3217/978-3-85125-260-6-152

[27] Fazel-Rezai R, Ramanna S. Brain signals: Feature extraction and classification using rough set methods. In: Ślęzak D, Yao J, Peters JF, Ziarko W, Hu X, editors. Rough Sets, Fuzzy Sets, Data Mining, and Granular Computing. RSFDGrC 2005. Vol. 3642. Lecture Notes in Computer Science. Berlin, Heidelberg: Springer; 2005. pp. 709-718

[28] Kleih SC, Kaufmann T, Zickler C, Halder S, Leotta F, Cincotti F, et al. Out of the frying pan into the fire-the P300-based BCI faces real-world challenges. Progress in Brain Research. 2011;**194**:27-46

[29] Mak JN, Arbel Y, Minett JW, McCane LM, Yuksel B, Ryan D, et al. Optimizing the P300-based brain-computer interface: Current status, limitations and future directions. Journal of Neural Engineering. 2011;**8**(2):25003

[30] Farwell LA, Donchin E. Talking off the top of your head: Toward a mental prosthesis utilizing event-related brain potentials. Electroencephalography and Clinical Neurophysiology. 1988;**70**(6):510-523

[31] Donchin E, Spencer KM, Wijesinghe R. The mental prosthesis: Assessing the speed of a P300-based brain- computer interface. IEEE Transactions on Rehabilitation Engineering. 2000;**8**(2):174-179

[32] Xu N, Gao X, Hong B, Miao X, Gao S, Yang F. BCI competition 2003 - Data set IIb: Enhancing P300 wave detection using ICA-based subspace projections for BCI applications. IEEE Transactions on Biomedical Engineering. 2004;**51**(6):1067-1072

[33] Serby H, Yom-Tov E, Inbar GF. An improved P300-based brain-computer interface. IEEE Transactions on Neural Systems and Rehabilitation Engineering. 2005;**13**(1):89-98

[34] McGillem CD, Aunon JI. Measurements of signal components in single visually evoked brain potentials. IEEE Transactions on Biomedical Engineering. 1977;**24**(3):232-241

[35] Guger C, Daban S, Sellers E, Holzner C, Krausz G, Carabalona R, et al. How many people are able to control a P300-based brain-computer interface (BCI)? Neuroscience Letters. 2009;**462**(1):94-98

[36] Interface BC, Thulasidas M, Guan C, Member S, Wu J, Speller AP. Robust classification of EEG signal for. IEEE Transactions on Neural. 2006;**14**(1):24-29

[37] Mirghasemi H, Shamsollahi MB, Fazel-Rezai R. Assessment of preprocessing on classifiers used in the P300 speller paradigm. In: Annual International Conference of the IEEE Engineering in Medicine and Biology Society (EMBS), New York, NY. 2006. p. 1319-22. DOI: 10.1109/IEMBS.2006.259520

[38] Huang A, Zhou W. BLDA approach for classifying P300 potential. In: Peng Y, Weng X, editors. 7th Asian-Pacific Conference on Medical and Biological Engineering. IFMBE Proceedings. Vol. 19. Berlin, Heidelberg: Springer; 2008. pp. 341-343

[39] Mirghasemi H, Fazel-Rezai R, Shamsollahi MB. Analysis of P300 classifiers in brain computer interface speller. In: Annual International Conference of the IEEE Engineering in Medicine and Biology Society, New York, NY. 2006. pp. 6205-6208. DOI: 10.1109/IEMBS.2006.259521

[40] Dal Seno B, Matteucci M, Mainardi L. Online detection of P300 and error potentials in a BCI speller. In: Computational Intelligence and Neuroscience. Vol. 2010. Article ID 307254, 5 p, 2010. DOI: 10.1155/2010/307254

[41] Pires G, Castelo-Branco M, Nunes U. Visual P300-based BCI to steer a wheelchair: A Bayesian approach. In: Annual International Conference of IEEE Engineering in Medicine and Biology Society, Vancouver, BC. 2008. pp. 658-661. DOI: 10.1109/IEMBS.2008.4649238

[42] Hoffmann U, Yazdani A, Vesin JM, Ebrahimi T. Bayesian feature selection applied in a P300 brain-computer interface. In: European Signal Processing Conference, Lausanne. 2008. pp. 1-5

[43] Rakotomamonjy A, Guigue V. BCI competition III: Dataset II- Ensemble of SVMs for BCI P300 speller. IEEE Transactions on Biomedical Engineering. 2008;**55**(3):1147-1154

[44] Garrett D, Peterson DA. Comparison of linear, nonlinear, and feature selection methods for EEG signal classification. IEEE Transactions on Neural Systems and Rehabilitation Engineering. 2003;**11**(2):141-144

[45] Kaper M, Meinicke P, Grossekathoefer U. BCI competition 2003-data set IIb – support vector machines for the P300 speller paradigm. IEEE Transactions on Biomedical Engineering. 2004;**51**(6):1073-1076. DOI: 10.1109/TBME.2004.826698

[46] Vaughan TM, McFarland DJ, Schalk G, SarnackiWA, Krusienski DJ, Sellers E W, Wolpaw JR. The wadsworth BCI research and development program. IEEE Transactions on Neural Systems and Rehabilitation Engineering. 2006;**14**(2):229-233

[47] Sellers EW, Vaughan TM, Wolpaw JR. A brain-computer interface for long-term independent home use. Amyotrophic Lateral Sclerosis. 2010;**11**(5):449-455

[48] Allison BZ, Pineda JA. ERPs evoked by different matrix sizes: Implications for a brain computer interface (BCI) system. IEEE Transactions on Neural Systems and Rehabilitation Engineering. 2003;**11**(2):110-113

[49] Fazel-Rezai R, Abhari K. A region-based P300 speller for brain-computer interface. Computers & Electrical Engineering. 2009;**34**(3):81-85

[50] Townsend G, LaPallo B, Boulay C. ScienceDirect – Clinical Neurophysiology : A Novel P300-based Brain-Computer Interface Stimulus Presentation Paradigm: Moving Beyond Rows and Columns [Internet]. Clinical. 2010. Available from: http://www.sciencedirect.com/science/article/pii/S1388245710000738 [Accessed: 31 January 2017]

[51] Kaufmann T, Schulz SM, Grünzinger C, Kübler A. Flashing characters with famous faces improves ERP-based brain-computer interface performance. Journal of Neural Engineering. 2011;**8**(5):56016

[52] Jin J, Allison BZ, Wang X, Neuper C. A combined brain-computer interface based on P300 potentials and motion-onset visual evoked potentials. Journal of Neuroscience Methods. 2012;**205**(2):265-276

[53] Alamdari N, Haider A, Arefin R, Verma AK, Tavakolian K, Fazel-Rezai R. A review of methods and applications of brain computer interface systems. In: IEEE International Conference on Electro Information Technology (EIT), Grand Forks, ND; 2016. pp. 0345-0350. DOI: 10.1109/EIT.2016.7535263

[54] Fazel-Rezai R, Ahmad, Waqas A, Guger C, Edlinger G, Krausz G. Recent advances in brain-computer interface systems. In: Fazel-Rezai R, editor. Recent Advances in Brain-Computer Interface Systems. Rijeka, Croatia: InTech; 2011. pp. 1-24

[55] Mugler E, Bensch M, Halder S, Rosenstiel W, Birbaumer N, Kübler A. Control of an internet browser using the P300 event-related potential. Control. 2008;**10**(1):56-63

[56] Münßinger JI, Halder S, Kleih SC, Furdea A, Raco V, Hösle A, et al. Brain painting: First evaluation of a new brain-computer interface application with ALS-patients and healthy volunteers. Frontiers in Neuroscience. 2010:182-195. DOI: 10.3389/fnins.2010.00182

[57] Boland D, Quek M, Tangermann M, Williamson J, Murray-Smith R. Using simulated input into brain-computer interfaces for user-centred design. International Journal of Bioelectromagnetism. 2011;**13**(2):86-87

[58] Finke A, Lenhardt A, Ritter H. The MindGame: A P300-based brain-computer interface game. Neural Networks. 2009;**22**(9):1329-1333

[59] Mühl C, Gürkök H, Plass-Oude Bos D, Thurlings MEM, Scherffig L, Duvinage M, et al. Bacteria Hunt: A multimodal, multiparadigm BCI game. Enterface2009. 2009;**1**:1-22

[60] Congedo M, Goyat M, Tarrin N, Varnet. L, Rivet B, Ionescu G, et al. "Brain Invaders": A prototype of an open-source P300-based video game working with the OpenViBE

platform. In: 5th International Brain-Computer Interface Conference 2011 (BCI 2011), Graz, Austria. 2011. pp. 280-283

[61] Farwell LA, Donchin E. The truth will out: Interrogative polygraphy ("lie detection") with even-related brain potentials. Psychophysiology. 1991;**28**(5):531-547

[62] Smith SS, Agent S, Farwell L a, Smith SS. Using brain MERMER testing to detect knowledge despite efforts to conceal. Journal of Forensic Sciences. 2001;**46**(1):135-143

[63] Farwell LA. Method and apparatus for truth detection. US Patent 5406956; 1995

[64] Yuan P, Wang Y, Gao X, Jung T-P, Gao S. A collaborative Brain-Computer interface for accelerating human decision making. In: Stephanidis C, Antona M, editors. Universal Access in Human-Computer Interaction. Design Methods, Tools, and Interaction Techniques for Inclusion. UAHCI 2013. Vol. 8009. Lecture Notes in Computer Science. Berlin, Heidelberg: Springer; 2013. pp. 672-681. DOI: 10.1007/978-3-642-39188-0_72

[65] Wang H, Chang W, Zhang C. Functional brain network and multichannel analysis for the P300-based brain computer interface system of lying detection. Expert Systems with Applications. 2016;**53**:117-128

[66] Wolpaw JR, McFarland DJ, Neat GW, Forneris CA. An EEG-based brain-computer interface for cursor control. Electroencephalography and Clinical Neurophysiology. 1991;**78**(3):252-259

[67] Fazel-Rezai R, Pauls M, Slawinski D. A low-cost biomedical signal transceiver based on a bluetooth wireless system. In: Annual International Conference of the IEEE Engineering in Medicine and Biology – Proceedings; 2007. pp. 5711-5714

[68] Lewand R. Cryptological Mathematics. Washington, DC: Mathematical Association of America; 2000. ISBN: 0883857197

[69] Fazel-Rezai R, Abhari K. A comparison between a matrix-based and a region-based P300 speller paradigms for brain-computer interface. Conference IEEE Engineering in Medicine and Biology Society. 2008;**2008**:1147-1150

[70] Amiri S, Fazel-Rezai R, Asadpour V. A review of hybrid brain-computer interface systems. In: Advances in Human-Computer Interaction. Vol. 2013. Article ID 187024, 2013. pp. 1-8. DOI: 10.1155/2013/187024

[71] Amiri S, Rabbi A, Azinfar L, Fazel-Rezai R, Asadpour V. A review of P300, SSVEP, and hybrid P300/SSVEP brain-computer interface systems. Brain-Computer Interface System – Recent Progress and Future Prospects. 2013;**2013**:1-8

[72] Haider A, Cosatto B, Alam M, Tavakolian K, Fazel-Rezai R. A new Region-based BCI speller design using steady state visual evoked potentials. In: 6th International Brain-Computer Interface Meeting, organized by the BCI Society; Pacific Gove, CA. 2016. p. 1. Verlag der TU Graz, Graz University of Technology. DOI: 10.3217/978-3-85125-467-9-1

[73] Fazel-Rezai R. Human error in P300 speller paradigm for brain-computer interface. In: Annual International Conference of the IEEE Engineering in Medicine and Biology Society, Lyon. 2007. pp. 2516-2519. DOI: 10.1109/IEMBS.2007.4352840

[74] Stanney K, Schmorrow D, Johnston M, Fuchs S, Jones D, Hale KS, et al. Augmented cognition: An overview. Reviews of Human Factors and Ergonomics. 2009;**5**(1):195-224

Chapter 3

Motor Evoked Potentials in Supratentorial Glioma Surgery

Stefan Grossauer, Yaroslav Parpaley and Katharina Koeck

Additional information is available at the end of the chapter

http://dx.doi.org/10.5772/intechopen.70040

Abstract

Primary brain tumors, that is gliomas, are frequently located close to or within functional motor areas and motor tracts and therefore represent a major neurosurgical challenge. Preservation of the patients' motor functions, while achieving a maximum resection of tumor, can be only achieved by monitoring and locating motor areas and motor tracts intraoperatively. The intraoperative use of motor evoked potentials (MEPs) represents the current gold standard to do so. However, intraoperative MEP monitoring and mapping can be quite challenging and require a profound knowledge of the MEP technique, brain anatomy and physiology and anesthesia. In this chapter, a systematic review of PubMed listed literature on MEP monitoring and mapping in glioma surgery is presented. The benefits, limitations, technical pearls and pitfalls are discussed from the perspective of an experienced neurosurgical/neurophysiological team.

Keywords: intraoperative neurophysiological monitoring, glioma surgery, motor evoked potentials, intraoperative motor mapping

1. Introduction

Primary brain tumors, that is gliomas, which are frequently located close to or within functional motor areas and motor tracts of the brain [1] represent a major neurosurgical challenge [2]. Surgery-related neurological deficits often arise from direct damage to the cortical or subcortical structures or from ischemia [3, 4]. Preservation of the patients' motor functions, while achieving a maximum resection of tumor, can be only achieved by monitoring and locating motor areas and motor tracts intraoperatively [5, 6]. Therefore, it is nowadays agreed within

INTECH
open science | open minds

© 2017 The Author(s). Licensee InTech. This chapter is distributed under the terms of the Creative Commons Attribution License (http://creativecommons.org/licenses/by/3.0), which permits unrestricted use, distribution, and reproduction in any medium, provided the original work is properly cited. (cc) BY

the neurosurgical community that the intraoperative use of motor evoked potentials (MEPs) represents the current gold standard to monitor and map the primary motor cortex and the corticospinal tract (CST).

As early as in the 1980s, Merton and Morton discovered that a high-voltage single electrical stimulus applied over the skull could activate the motor cortex and the subcortical motor pathways and consequently generate MEPs, which could be easily recorded from the limb muscles [7]. This finding was then exploited for the development of intraoperative neuromonitoring techniques that involve MEPs induced via both direct cortical and transcranial cortical electrical stimulation and direct subcortical white matter stimulation [7–9]. Since then, the intraoperative use of MEPs has substantially contributed to improved functional outcomes of glioma patients [10] as well as pushing the boundaries of surgery for lesions considered inoperable in the pre-MEP era [6].

However, intraoperative MEP monitoring and mapping can be quite challenging and require a profound knowledge of the MEP technique, brain anatomy and physiology and anesthesia.

Therefore, it is usually accomplished with the combined efforts of a multidisciplinary team of neurosurgeons, neuroradiologists, neuroanesthesiologists and intraoperative neurophysiologists, who together contribute to prevent neurological damage in glioma patients. Given that each tumor induces specific modifications in the brains' functional network, surgery must be tailored according to functional and anatomical boundaries of each patient individually [11].

2. Technique

The intraoperative use of MEPs encompasses both monitoring and mapping techniques. Whereas monitoring techniques continuously assess the functional integrity of the CST by repetitively testing MEPs frequently during surgery, MEP mapping instead is designed to identify the primary motor cortex and the subcortical CST within ambiguous neural tissue at appropriate surgical stages. For example, mapping the exposed cortex during the early stages of surgery enables to localize or, if not exposed, rules out the primary motor cortex in the surgical field before the cortex is incised and therefore harmed. In the later stages of surgery, it is often critical to localize the CST within the subcortical white matter during tumor resection, therefore it is localized using subcortical MEP mapping. Conversely, MEPs after transcranial and/or direct cortical stimulation are frequently elicited and recorded during all stages of surgery to proof the integrity of the CST and recognize impending damage to it when it is still potentially reversible. MEP monitoring therefore enables the surgeon to adjust the surgical manipulation early enough to prevent permanent damage to the cerebral motor system.

Figure 1 shows the stimulation and recording sites for MEPs as used at our and most other institutions during surgeries for supratentorial gliomas within or near to the primary motor cortex and/or CST. It also illustrates the differences in technique for the different stimulation sites which are transcranial MEPs (t-MEPs), direct cortical MEPs (dc-MEPs) and subcortical MEPs (sc-MEPs). **Tables 1** and **2** summarize the stimulation and recording parameters used at our institution.

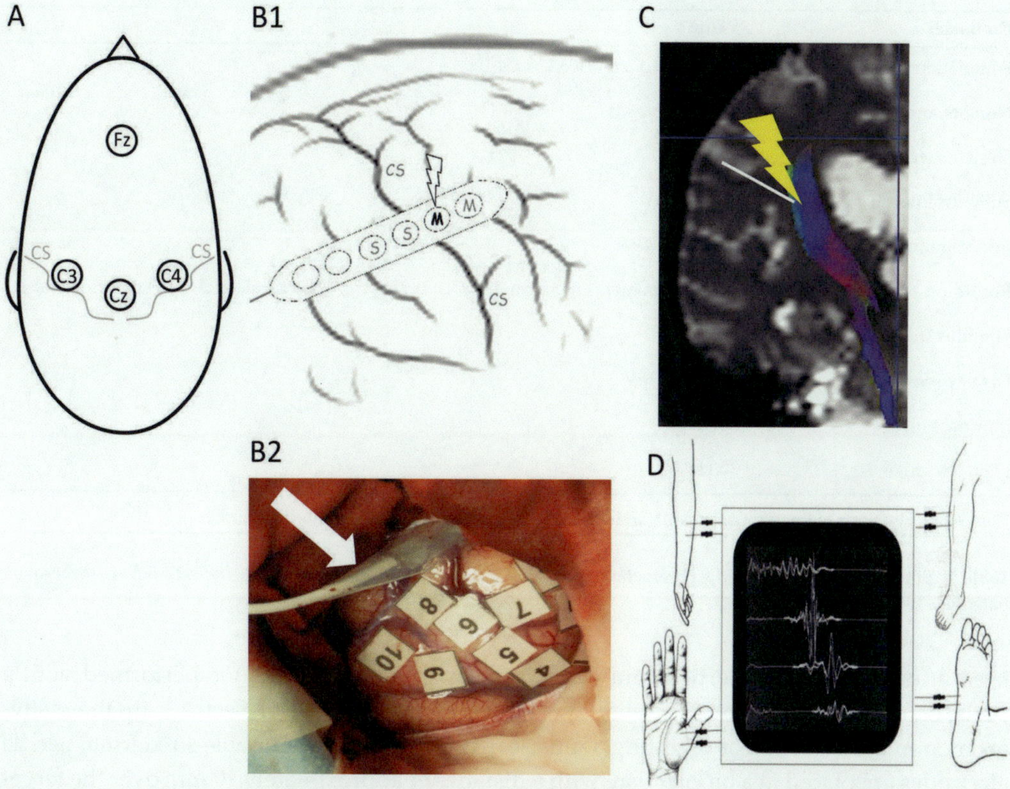

Figure 1. Stimulation and recording sites for motor evoked potentials (MEPs) used during surgeries for supratentorial gliomas. Usually, different stimulation sites are employed during surgery to elicit different types of MEPs that give distinct information intraoperatively. A shows the electrode montage for transcranial MEPs using corkscrew scalp electrodes in relation to the central sulcus (CS). The active electrodes (C3, C4, Cz) are located over the precentral gyrus using anatomical surface landmarks according to the international 10–20 EEG system (in black circles). B1 shows the stimulation site for direct cortical MEPs using a strip electrode with six contacts placed on the primary sensory cortex (S) and primary motor cortex (M). One contact over the primary motor cortex is chosen as active contact (bold M and flash) to deliver currents with comparably low intensity directly to the motor cortex. B2 shows the corresponding intraoperative photograph to scheme B1 taken after craniotomy, dura opening, placement of the strip electrode (white arrow) beneath the dura onto the cortex. The strip electrode is placed under the craniotomy margin away from the surgical field to avoid interference with surgery. Sterile numbers are placed on the cortex for the documentation of the functional mapping results. C shows a magnetic resonance imaging illustrating the corticospinal tract using diffusion tensor imaging technique. For subcortical MEP mapping, the subcortical white matter and the corticospinal tract are stimulated using a stimulation probe or a surgical suction tube with an electrode tip (white bar with yellow flash). D shows the upper and lower extremity target muscles used for MEP recorded during supratentorial glioma surgery. The MEPs are displayed on a screen and their amplitudes and latencies are analyzed continuously during surgery. CS=central sulcus, M=contact on the primary motor cortex, S=contact on the primary sensory cortex.

MEP monitoring is routinely performed by a dedicated team in all glioma cases of presumed tumor location close to the primary motor cortex or the corticospinal tract. This is achieved either by transcranial electrical stimulation with cork screw scalp electrodes (t-MEPs) (**Figure 1A**) or by direct cortical stimulation (dc-MEPs) via a cortical strip electrode placed on the precentral gyrus (**Figure 1B1**). In both cases, a constant current anodal stimulation, train-of-5 stimuli with an interstimulus interval of 4.0 ms and an impulse width of 500 μs are

Parameter	Value
Stimulation sites	C3/4; C4/3; Cz/Fz: C3/Cz; C4/Cz
Number and form of stimuli	Train of five
Stimulus lengths	500 μs
Stimulus intensity	30–220 mA
Interstimulus interval	4 ms
Recording sites	ABP contralateral, forearm flexor contralateral, TA contralateral, ABH contralateral
Stimulus frequency	0.5 Hz
Display sensitivity	200 μV–2 mV
Low pass filter	3000 Hz
High pass filter	30 Hz

Note: ABH, abductor hallucis muscle; ABP, abductor pollicis muscle; TA, tibialis anterior muscle.

Table 1. Stimulation and recording parameters used for intraoperative monitoring with transcranial motor evoked potentials (t-MEPs).

used at our institution. Cortical and subcortical mapping (**Figure 1C**) is performed with a probe or, even more comfortably, with a surgical suction tube with an electrode tip that delivers monopolar current. For detecting a muscle, MEPs 27-gauge disposable subdermal needle electrodes are placed in a bipolar way with a distance of approximately 10 mm over the target muscles of the contralateral side of the tumor. We use the abductor pollicis brevis, the flexor carpi radialis for the upper extremity and the anterior tibial muscle and the abductor hallucis for the lower extremity.

Parameter	Value
Stimulation sites	Primary motor cortex, subcortical white matter
Number and form of stimuli	Train of five
Stimulus lengths	500 μs
Stimulus intensity	3–20 mA
Interstimulus interval	4 ms
Recording sites	ABP contralateral, forearm flexor contralateral, TA contralateral, ABH contralateral
Stimulus frequency	0.5 Hz
Display sensitivity	200 μV–2 mV
Low pass filter	3000 Hz
High pass filter	30 Hz

Note: ABH, abductor hallucis muscle; ABP, abductor pollicis muscle; TA, tibialis anterior muscle.

Table 2. Stimulation and recording parameters used for intraoperative monitoring with direct cortical motor evoked potentials (dc-MEPs), as well as subcortical motor evoked potentials (sc-MEPs) used for subcortical mapping.

Transcranial electrical stimulation is applied using corkscrew electrodes which are screwed into the scalp and therefore guarantee low impedances. We routinely place initially four electrodes at C3, C4, Cz and Fz and try to elicit muscle MEPs starting with a low stimulus intensity of 30 mA from a C3/Cz montage for left-sided tumors or a C4/Cz montage for right-sided tumors. Stimulus intensity is then increased in 5 mA steps until stable recordings from all muscles can be obtained (see **Tables 1** and **2** for target muscles). In the case of high stimulus intensities (above 150 mA) or vigorous muscle twitching, cork screw electrodes are added at C1, C2 and Cz+6 and stimulation is repeated with a C1/2, C2/1 or a Cz/Cz+6 montage again starting with low intensities that are incrementally increased. Using different montages of stimulating electrodes provides flexibility to optimize elicitation of muscle MEPs and avoiding muscle twitching which can interfere with surgery. Extremely high stimulus intensities are generally avoided because this might activate the corticospinal tract (CST) deep in the brain distal to the tumor and may therefore produce false-negative results [7, 12].

Because the intensity needed for dc-MEPs (**Table 2**) is 10 times lower than for t-MEPs (**Table 1**), dc-MEPs are preferred over t-MEPs in all cases, where the surgical approach allows access to the primary motor cortex. Even if the motor cortex itself is not exposed, a six to eight contact strip electrode for direct cortical stimulation is slid underneath the dura and oriented perpendicularly to the assumed central sulcus (**Figure 1B1/2**). Somatosensory evoked potentials (SEP) phase reversal can be used as a help to identify the central sulcus.

Then the first electrode in front of the sulcus is usually used as the stimulating anodal contact and the strip electrode is kept in place by a compress and by clamping it subdurally [7]. Usually, the cork screw electrode mounted at Fz serves as the cathodal pole.

Stimulation intensity begins with 5 mA and is increased continuously in steps of 2 mA until stable MEPs can repeatedly be obtained from all target muscles. Amplitudes are then evaluated by a trained intraoperative neurophysiologist measuring peak-to-peak differences as well as the latencies defined as the time span from start of the stimulation to the first assessable amplitude [13]. After this, a baseline is set and continuous monitoring is performed throughout the whole operation with an interval of at least 120 s. In stages of surgery where the CST is particularly endangered, the intervals are shortened to 10 s.

Monopolar subcortical stimulation to elicit sc-MEPs represents the gold standard for functional localization of the CST [14]. Its technique is described in **Figure 1**. It is used during resection of gliomas when closely approaching the CST in the deep white matter or to identify the primary motor cortex. When employed to evaluate the distance of the stimulation site to the CST, an initial stimulus intensity of 15 mA is used initially. When MEPs can be obtained at 15 mA after stimulating the wall of the resection cavity, stimulation intensities are gradually decreased in 5 mA steps until MEP responses can be obtained with a stimulus intensity as low as 5 mA. Given that this indicates a close proximity of around 5 mm to the CST, resection is usually stopped there.

3. Warning criteria and functional outcome

Warning criteria for MEP monitoring and mapping during glioma surgery must be carefully defined to be able to warn the surgeon early enough to prevent permanent damage to the patients' motor system, while not impeding tumor removal to early. This requires a well-balanced approach between the two rivaling goals, that is tumor removal and functional integrity. This means that if the warning criteria are too cautiously defined, it would lead to many false-positive test results with negative impact on the extent of tumor resection in functionally intact patients. On the other hand, less restrictive warning criteria would lead to more false-negative test results putting more patients at risk of impaired motor functions while enhancing the extent of tumor resection.

Whereas there is rather high proportion of false-positive results, false-negative results are rarely reported and can typically be explained on the basis of basic errors in technique or interpretation [15].

The optimal MEP technique and warning criteria would lead to true test results only, while avoiding false-negative results with patients sustaining new motor deficits and also avoiding false-positive results indicating motor deficits that these patients will actually not exhibit postoperatively.

The most employed criterion for MEP monitoring is the amplitude criterion. This means that the stimulus intensity which is able to elicit stable MEPs in all target muscles is set at the beginning of surgery and kept the same throughout the whole operation. From there on, a drop of MEP amplitude of 50% or more compared with the baseline amplitude results in a warning. To a lesser extent, the threshold criterion is also used by some groups in addition to the amplitude criterion. This means that, in cases where an increase in stimulus intensity of 20% or more is needed to elicit MEPs, the surgeon is warned.

Concerning the stimulation site, most groups rely on dc-MEP monitoring; in all cases the surgical approach enables them to do so rather than on t-MEPs, for the reasons mentioned above.

However, a report by Lee et al. [16] shows that relying on t-MEP monitoring for supratentorial lesion surgery leads to similar results concerning test-accuracy and functional outcomes compared with reports from groups heavily relying on dc-MEP monitoring.

Table 3 summarizes the warning criteria and functional outcomes of surgical series reported in the literature. The list also includes a series of 95 patients from the authors of this chapter who underwent surgery for perirolandic gliomas with the aid of MEP neuromonitoring and mapping.

A new approach to the threshold criterion was recently described in a publication by Abboud et al. [17]. The authors evaluated the accuracy of changes in threshold level involving contra- and ipsilateral MEPs, contrary to the usual approach to compare the changes of ipsilateral MEPs to the baseline recording. Ninety-three patients underwent t-MEP monitoring during resection of gliomas located close to central motor pathways but not involving the primary motor cortex. An increase in the threshold level on the contralateral side of more than 20% beyond the percentage increase on the ipsilateral side was considered a significant alteration.

Interestingly none of their patients without a significant threshold increase exhibited a new motor deficit postoperatively, whereas all 13 patients with a significant MEP alteration

Author/year	MEP techniques	Warning criteria	New motor deficits	New motor deficits undetected by MEP monitoring (transient and permanent)
Authors of this chapter/2017	t-MEPs, dc-MEPs	Amplitude<50% and/or threshold increase of 20%	23% transient 6% permanent	2.1%
Abboud/2016 [17]	t-MEPs bilateral	Threshold increase >20%	14% (transient and permanent)	0%
Obermueller/2015 [18]	dc-MEPs	Amplitude<50%	19% transient 14% permanent	N/A
Lee/2014 [16]	t-MEPs	Amplitude<50%	10% transient 7% permanent	13%
Krieg/2013 [19]	dc-MEPs	Amplitude<50%	37% transient 11% permanent	N/A
Krammer/2009 [20]	t-MEPs	Amplitude<50% and/or threshold increase of 20%	15% (transient and permanent)	1.6%
Neuloh/2007 [21]	t-MEPs, dc-MEPs	Amplitude<50%	21% transient 10% permanent	2.7%
Sala/2003 [22]	t-MEPs, dc-MEPs	Amplitude<50%	24% transient	N/A

The list also includes a series of 95 patients from the authors of this chapter who underwent surgery for perirolandic gliomas with the aid of MEP neuromonitoring and mapping.

dc-MEP, direct cortical motor evoked potentials; N/A, data not available; t-MEPs, transcranial motor evoked potentials.

Table 3. MEP methods, the warning criteria and functional outcomes in neurosurgical series reported in the literature.

exhibited a new motor deficit. Hence, this method resulted in a 100% sensitivity and specificity. This is the first surgical series showing such a high accuracy of MEP monitoring. Although no new motor deficit was undetected with this method, 14% of their patients sustained new motor deficits, nevertheless. This signifies the importance of not just a high accuracy but also a timely warning to prevent neurological deficit.

4. Illustrative case

A 26-year-old male patient, who has undergone surgery for a left frontal low-grade astrocytoma 2 years earlier, exhibited a significant recurrent tumor on the latest magnetic resonance images (MRI) posteriorly to the old resection cavity as shown in **Figure 2A**. Preoperative functional imaging using MRI tractography (**Figure 2B**) and functional MRI (**Figure 2C**) proved the tumor's close relationship to the CST and the primary motor cortex. Therefore, it was decided to employ MEP monitoring and mapping for the planned resection of this astrocytoma.

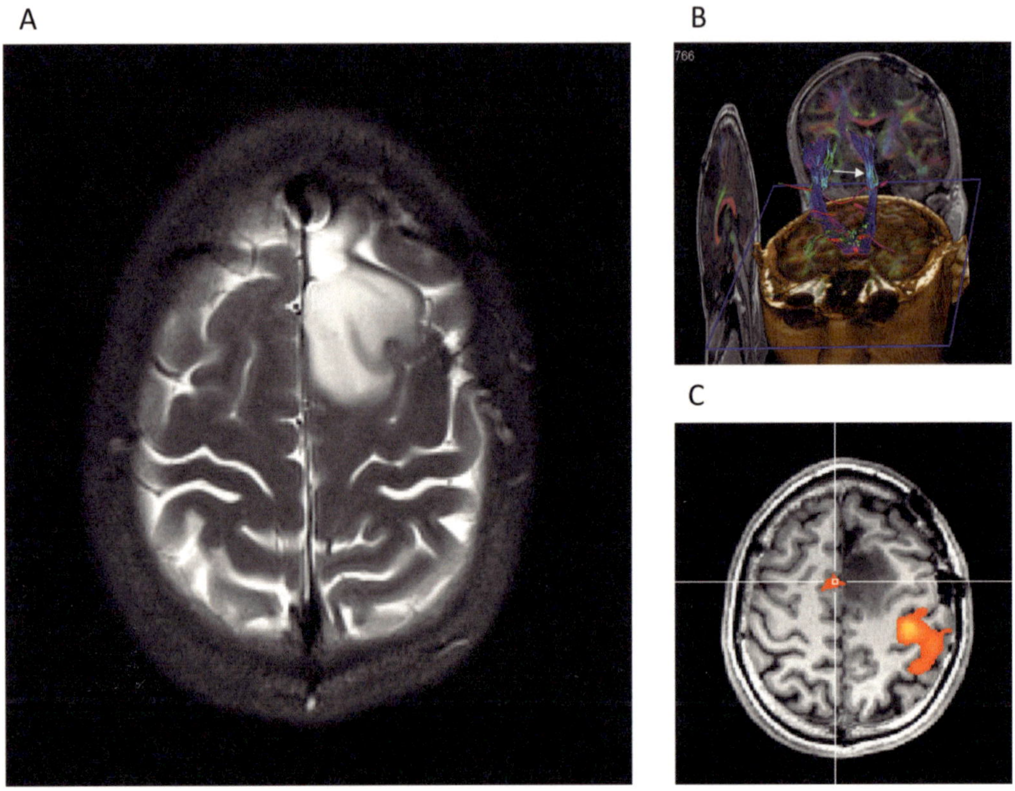

Figure 2. Preoperative images from a patient harboring a recurrent left frontal low-grade astrocytoma. Preoperative work up included T2-weighted magnetic resonance imaging (MRI) (A) showing the close relationship of the posterior tumor margin to the precentral gyrus, as well as MRI tractography and (B) showing the left-sided corticospinal tract displaced posteriorly (white arrow) by the tumor and a functional MRI of a right-sided finger-tapping paradigm showing activation in the area of the precentral gyrus' hand knob.

After positioning the patient in total intravenous general anesthesia, subcutaneous needle electrodes were placed over the target muscles of the right upper and lower extremity. Given that a surgical approach was planned that would allow the placement of a strip electrode over the primary motor cortex to obtain dc-MEPs, cork screw electrodes were only placed as a back-up in case t-MEPs have to be recorded. All electrodes were then connected to the neuromonitoring device and the electrode impedances were checked. Then the surgical approach was established by performing a left frontal paramedian craniotomy that also exposed the medial part of the left precentral gyrus. The dura was opened and the precentral gyrus was identified using monopolar direct cortical stimulation with an intensity of 10 mA. The strip electrode was placed subdurally perpendicular to the central sulcus in a direction away from the surgical field to avoid interference with the surgical procedure. Direct cortical stimulation for dc-MEPs was started with an intensity of 5 mA and incrementally increased up to the supramaximal intensity of 13mA. Stable MEPs could be obtained from all contralateral target muscles. Further increase of the stimulus intensity at this point did not further increase the MEP amplitudes so that the baseline was set and measurements were taken. Monopolar subcortical stimulation was set up then using the surgical suction tip for anodal subcortical stimulation throughout the whole operation, starting at an initial intensity of 15 mA. Corticotomy was performed and the anterior part of the tumor-harboring left superior frontal gyrus was resected without any significant changes of MEP amplitudes or latencies. As approaching the subcortical white matter just anterior to the precentral gyrus, MEPs in all upper extremity muscles could be elicited with the suction tip stimulation. The surgeon was alerted and the stimulus intensity was lowered to 10 mA which did not elicit any potentials. Cautious further resection of tumor at its posterior aspect suddenly led to a significant amplitude decline of the MEPs recorded from the abductor pollicis (ABP) of more than 50% compared with the baseline (**Figure 3**). The surgeon was alerted, surgery was stopped, the spatula was removed and the resection cavity was irrigated with warm Ringers' solution. After 12 min when the MEPs have recovered, a small remnant tumor was removed. Just before wound closure, the last intraoperative recordings showed no significant change compared to the baseline recording, indicating normal postoperative motor functions. Surgery was completed uneventfully.

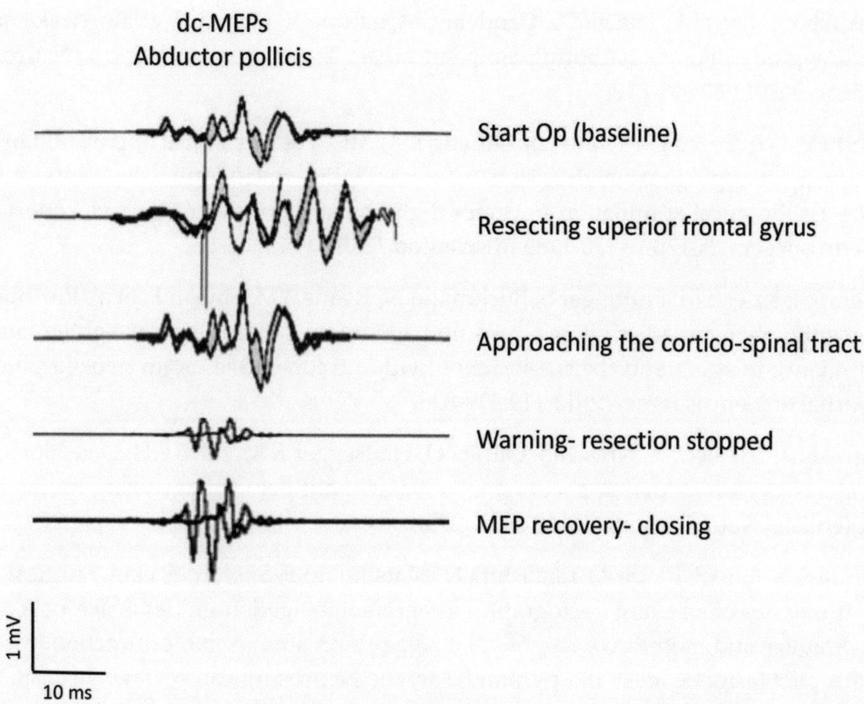

Figure 3. Intraoperative direct cortical motor evoked potentials (dc-MEPs) taken from the right abductor pollicis of a patient undergoing surgery for a left frontal recurrent astrocytoma. Baseline MEPs were obtained just after the placement of the strip electrode over the precentral gyrus. There were no significant changes of MEP amplitudes or latencies until subcortical resection was carried out in close proximity to the corticospinal tract at the posterior aspect of the tumor. Warning was given by the intraoperative neurophysiologist when a drop of MEP amplitude of more than 50% compared with the baseline recording was noted. Surgery was stopped, the spatula was removed from the resection cavity and the operating field was irrigated with warm Ringers' solution. After 12 min, the MEPs were recovered and a small remnant tumor was removed. Just before wound closure, the last intraoperative recordings showed no significant change compared to the baseline recording indicating normal postoperative motor functions.

In the immediate postoperative course, the patient exhibited a right-sided hemiplegia which recovered fully within 2 weeks and therefore proved to represent a supplemental motor area syndrome rather than a damage of the CST.

Author details

Stefan Grossauer*, Yaroslav Parpaley and Katharina Koeck

*Address all correspondence to: st.grossauer@gmail.com

Department of Neurosurgery, Ruhr University Bochum, Germany

References

[1] Sarubbo S, Latini F, Panajia A, Candela C, Quatrale R, Milani P, et al. Awake surgery in low-grade gliomas harboring eloquent areas: 3-year mean follow-up. Neurological Sciences. 2011;**32**:801-810

[2] Zhu FP, Wu JS, Song YY, Yao CJ, Zhuang DX, Xu G, et al. Clinical application of motor pathway mapping using diffusion tensor imaging tractography and intraoperative direct subcortical stimulation in cerebral glioma surgery: A prospective cohort study. Neurosurgery. 2012;**71**:1170-1183 [discussion 1183-1174]

[3] Gempt J, Krieg SM, Huttinger S, Buchmann N, Ryang YM, Shiban E, et al. Postoperative ischemic changes after glioma resection identified by diffusion-weighted magnetic resonance imaging and their association with intraoperative motor evoked potentials. Journal of Neurosurgery. 2013;**119**:829-836

[4] Grossauer S, Koeck K, Kraschl J, Olipitz O, Hausegger KA, Vince GH. Detection of cerebral vasospasm following aneurysmal subarachnoid hemorrhage using motor evoked potentials. Neurosurgery. 2016;**78**:265-273

[5] Ohue S, Kohno S, Inoue A, Yamashita D, Matsumoto S, Suehiro S, et al. Surgical results of tumor resection using tractography-integrated navigation-guided fence-post catheter techniques and motor-evoked potentials for preservation of motor function in patients with glioblastomas near the pyramidal tracts. Neurosurgical Review. 2015;**38**:293-306 [discussion 306-297]

[6] Schucht P, Seidel K, Beck J, Murek M, Jilch A, Wiest R, et al. Intraoperative monopolar mapping during 5-ALA-guided resections of glioblastomas adjacent to motor eloquent areas: Evaluation of resection rates and neurological outcome. Neurosurgical Focus. 2014;**37**:E16

[7] Sala F, Manganotti P, Grossauer S, Tramontanto V, Mazza C, Gerosa M. Intraoperative neurophysiology of the motor system in children: A tailored approach. Child's Nervous System. 2010;**26**:473-490

[8] Majchrzak K, Bobek-Billewicz B, Tymowski M, Adamczyk P, Majchrzak H, Ladzinski P. Surgical treatment of insular tumours with tractography, functional magnetic resonance imaging, transcranial electrical stimulation and direct subcortical stimulation support. Neurologia i Neurochirurgia Polska. 2011;**45**:351-362

[9] Sakurada K, Matsuda K, Funiu H, Kuge A, Takemura S, Sato S, et al. Usefulness of multimodal examination and intraoperative magnetic resonance imaging system in glioma surgery. Neurologia Medico-Chirurgica (Tokyo). 2012;**52**:553-557

[10] Yoshikawa K, Kajiwara K, Morioka J, Fujii M, Tanaka N, Fujisawa H, et al. Improvement of functional outcome after radical surgery in glioblastoma patients: The efficacy of a navigation-guided fence-post procedure and neurophysiological monitoring. Journal of Neuro-Oncology. 2006;**78**:91-97

[11] Bertani G, Fava E, Casaceli G, Carrabba G, Casarotti A, Papagno C, et al. Intraoperative mapping and monitoring of brain functions for the resection of low-grade gliomas: Technical considerations. Neurosurgical Focus. 2009;**27**:E4

[12] Li F, Deshaies EM, Allott G, Canute G, Gorji R. Direct cortical stimulation but not transcranial electrical stimulation motor evoked potentials detect brain ischemia during brain tumor resection. American Journal of Electroneurodiagnostic Technology. 2011;**51**:191-197

[13] Krieg SM, Shiban E, Droese D, Gempt J, Buchmann N, Pape H, et al. Predictive value and safety of intraoperative neurophysiological monitoring with motor evoked potentials in glioma surgery. Neurosurgery. 2012;**70**:1060-1070 [discussion 1070-1061]

[14] Ohue S, Kohno S, Inoue A, Yamashita D, Harada H, Kumon Y, et al. Accuracy of diffusion tensor magnetic resonance imaging-based tractography for surgery of gliomas near the pyramidal tract: A significant correlation between subcortical electrical stimulation and postoperative tractography. Neurosurgery. 2012;**70**:283-293 [discussion 294]

[15] Ritzl EK. Intraoperative neuromonitoring during glioma surgery: Bring in the expert neurophysiologists! Journal of Clinical Neurophysiology. 2012;**29**:151-153

[16] Lee JJ, Kim YI, Hong JT, Sung JH, Lee SW, Yang SH. Intraoperative monitoring of motor-evoked potentials for supratentorial tumor surgery. Journal of Korean Neurosurgical Society. 2014;**56**:98-102

[17] Abboud T, Schaper M, Duhrsen L, Schwarz C, Schmidt NO, Westphal M, et al. A novel threshold criterion in transcranial motor evoked potentials during surgery for gliomas close to the motor pathway. Journal of Neurosurgery. 2016;**125**:795-802

[18] Obermueller T, Schaeffner M, Shiban E, Droese D, Negwer C, Meyer B, et al. Intraoperative neuromonitoring for function-guided resection differs for supratentorial motor eloquent gliomas and metastases. BMC Neurology. 2015;**15**:211

[19] Krieg SM, Schnurbus L, Shiban E, Droese D, Obermueller T, Buchmann N, et al. Surgery of highly eloquent gliomas primarily assessed as non-resectable: Risks and benefits in a cohort study. BMC Cancer. 2013;**13**:51

[20] Krammer MJ, Wolf S, Schul DB, Gerstner W, Lumenta CB. Significance of intraoperative motor function monitoring using transcranial electrical motor evoked potentials (MEP) in patients with spinal and cranial lesions near the motor pathways. British Journal of Neurosurgery. 2009;**23**:48-55

[21] Neuloh G, Simon M, Schramm J. Stroke prevention during surgery for deep-seated gliomas. Neurophysiologie Clinique. 2007;**37**:383-389

[22] Sala F, Lanteri P. Brain surgery in motor areas: The invaluable assistance of intraoperative neurophysiological monitoring. Journal of Neurosurgical Sciences. 2003;**47**:79-88

Chapter 4

The Contingent Negative Variation: The Cumulative Curve Method Revisited

Daniel Dumalin

Additional information is available at the end of the chapter

http://dx.doi.org/10.5772/intechopen.69310

Abstract

The contingent negative variation (CNV) slow waves were elicited using a modified version of the standard paradigm matching the earlier work of Timsit-Berthier. Three parameters, the A3, A5 and post-imperative negative variation (PINV), are measured on four blocks of three to five trials and plotted into a cumulative curve. Five different types of cumulative curves are identified and used for further analysis of a clinical population. A literature review, applying the four-step approach for developing diagnostic tests in psychiatry by Boutros and colleagues is used to assess the current state of CNV as a clinical tool. Two clinical examples are used to illustrate that the cumulative curve reflects the current state of a mental disorder and that follow-up reflects the (un)favorable evolution. Clinical observations indicate that when taking into account the state of a mental disorder, the CNV has potential as a diagnostic aid and can play an active role in the therapeutic decision process.

Keywords: contingent negative variation, event-related potentials, slow cortical potentials, cumulative curve, post-imperative negative variation

1. Introduction

The earliest slow potentials (SPs) emerged from animal studies in the 1950s. Significant steps in man were only made, starting with Walter [1], when the switch to DC recording, or much longer time constants (TCs) were made, which was previously more common in animal work.

Since then, a flurry of studies has been published in an attempt to unravel the underlying processes concerned, variously characterized as arousal, activation, attention, priming, expectancy, motivation, etc. What all of the constructs had in common was an attempt to express the level of involvement existing between a subject and his test environment over a given

INTECH
open science | open minds

© 2017 The Author(s). Licensee InTech. This chapter is distributed under the terms of the Creative Commons Attribution License (http://creativecommons.org/licenses/by/3.0), which permits unrestricted use, distribution, and reproduction in any medium, provided the original work is properly cited.

period of (examination) time. From the 1980s onward, numerous studies on different patient populations were conducted in an effort to use the CNV as a diagnostic aid.

Despite all these efforts, a chasm seems to prevail between research findings and the experience of clinicians using the CNV as a clinical tool.

This is partly due to the fact that SPs offer its own range of artifactual problems and special technical considerations, such as the use of long time constants (TCs) and the difficulty to elicit reliable responses [2].

Another element that plays a role is addressed by the proposed four-step approach for developing diagnostic tests in psychiatry by Boutros et al. [3]. Most laboratory tests in psychiatry, such as the CNV, tend not to be developed into diagnostic tools. This can only lead to the above-mentioned disappointment and premature abandonment by early adopters or not even implementing a test by most practitioners.

For the clinician, it is important to know what the current status is of a test that shows promising results and see if progress is being made to develop the test into a diagnostic tool. A survey of literature (meta-analysis) should apply the four-step approach not only for clinicians, but also for researchers. Researchers should take count of the need for (simple and low cost) diagnostic tools for psychiatry. They also need to have a clear idea what the current status is of their research tool before progressing further. Is more evidence needed or are the requirements met to proceed to the next level?

The CNV was adopted in clinical practice in Belgium, thanks to the extensive work of Timsit-Berthier and some Belgium companies at the time that developed the equipment for clinical use. However, the clinical utility of the CNV is at present unknown and will be explored in this chapter.

From the above, it seems useful that there should be more interaction between the researcher and clinician, not only from the researcher to the clinician, but also the other way around. Clinicians have the advantage that in most cases they will adhere to a standardized procedure and will apply it for many more disorders than already explored in research. This could provide valuable feedback to researchers where they can best focus their attention first.

Clinicians also have a closer relationship with the test and their patients. They also will do repeated test. This has the potential for interesting observations that can lead to new discoveries that can benefit further development. This potential will be illustrated in this chapter with two clinical examples.

2. Methodology

2.1. Method

2.1.1. Stimuli and procedures

Each subject is presented with a series of paired signals of which the first S1 (warning signal) is a sound of 1 kHz with a duration of 0.5 s. The second sound S2 (imperative signal) is a continuous sound of 750 Hz with a duration of 5 s, which is delivered 1.5 s after S1. Both sounds are delivered through a loudspeaker (open field). The subject is instructed to keep the eyes closed and press a button (motor response (MR)) to stop the S2 sound at the point that it is clear for the subject that this is the target.

The signals are presented in series of three to five pairs and averaged in real time, four series in total. The duration of each trial is 8 s, with a prestimulus period of 1 s and post-S2 duration of 6 s. An intertrial randomized interval is used between 10 and 11.5 s (**Figure 1**).

Data from 20 EEG scalp electrodes placed according to the International 10–20 System of Electrode Placement with four artefact electrodes (one vertical and two horizontal eye movements, one EMG/ECG) are recorded with one ground electrode and two reference electrodes. Recording electrodes are referenced to linked ear electrodes and processed through analogue EEG amplifiers with a filter band-pass of 0.01–30 Hz with a total acquisition time of 8 s and digitalized with a sampling rate of 256 Hz.

A reject limit is set for each individual subject. This lets you automatically reject a transient high voltage pattern, such as eye-blink and eye-movement artifacts. The appropriate reject level is determined from monitoring the peak values of the first series of three to five pairs, which is discarded. The reject limit is usually set 10 μV higher than the maximum observed peak value. In most cases, this will be about 50 μV. On each trial, the reject limit is compared against the acquired signal amplitude in each channel. If 10 or more points in the input signal have an absolute amplitude value over the reject limit, the trial is rejected.

2.1.2. *Methodology background*

The Contingent Negative Variation (CNV) slow waves were elicited using a modified version of the standard paradigm of Timsit-Berthier, more closely matching in some respects the author's earlier work [4, 5] rather than the procedure described in her latest publication [6]. However, current method also deviates from Timsit-Berthier's procedures.

Current procedure aims to study the development of the CNV with the focus on the late components of the CNV.

The CNV reaches the maximum amplitude in normal adults in about 30 trials, although it can be fully developed in five to eight trials [7]. The assessment of the temporal evolution is done by

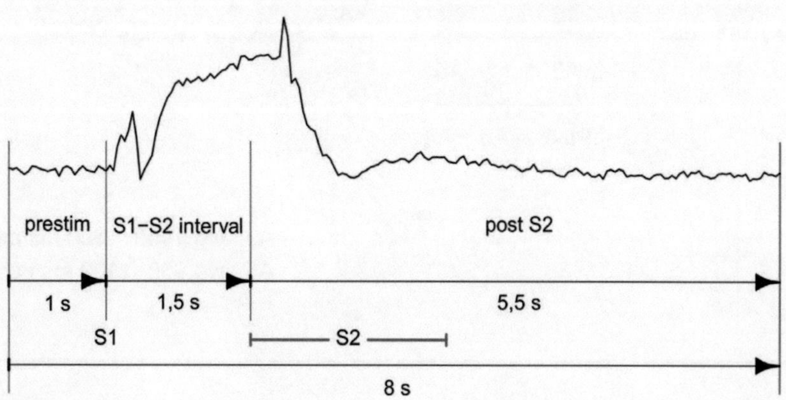

Figure 1. CNV cycle.

applying the cumulative curve method presented by Timsit-Berthier at EPIC IV in 1976 [5] and was never picked up again in later publications. Here we deviate from the presented procedure taking into account the number of trials at which the maximum CNV amplitude is reached. This limits the number of trials to 30 instead of 76, and the number averaged segments from 6 to 4.

The use of a constant-foreperiod reaction-time (S1-S2-MR) procedure is used to optimize the development of the CNV [7].

The S1–S2 time interval of 1.5 s is based on the earlier work of Timsit-Berthier in order to separate the late from the early component of the CNV. With a time interval of 1 s, as used in later publications, the early and late CNV components overlap.

2.2. Analysis

2.2.1. Parameters

2.2.1.1. A3, A5 and PINV

Three different parameters were measured on the CNV curve (**Figure 2**). First, the negative average amplitude of the late CNV was calculated between 1400 and 1600 ms (A3) after the start of S1 at Cz. Then, the positive average amplitude of the post-S2 amplitude was measured between 900 and 1100 ms (A5) after the start of S2 relative to average amplitude of the late CNV at Cz. Next, the difference between A5 and A3 was calculated. The difference is an expression of the post-imperative negative variation (PINV = A5 – A3).

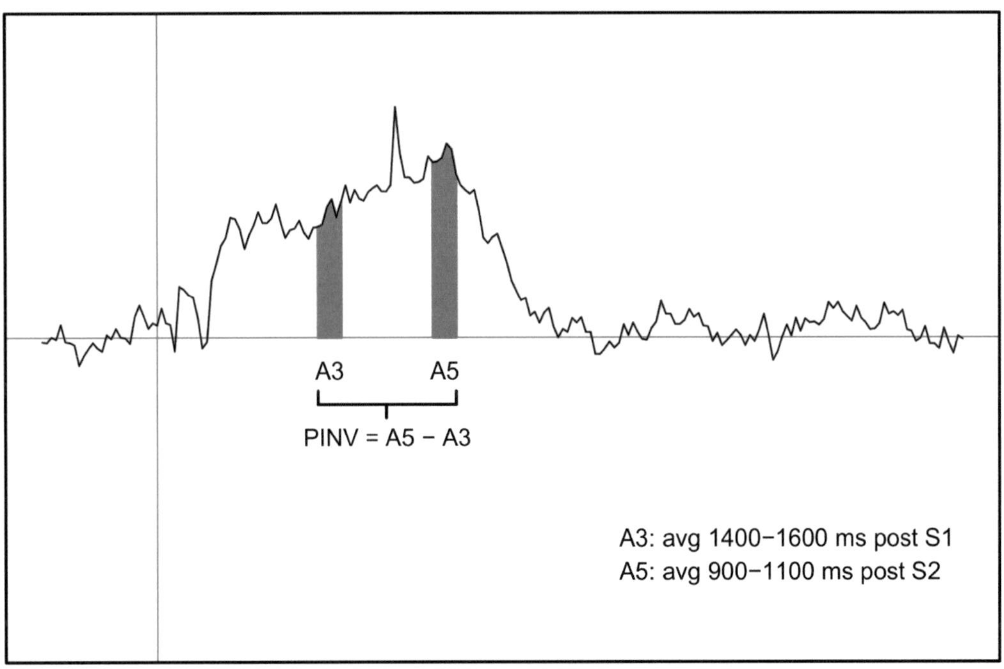

Figure 2. CNV parameters.

http://dx.doi.org/10.5772/intechopen.69310

These parameters are recorded for each of the four series of averaged trials.

2.2.1.2. Cumulative curve

Two separate amplitude segments are mapped in a cumulative histogram, with negative values above the *x*-axis. On each of the four sequential averages, we note the amplitude of A3 and PINV. On the horizontal axis, the first segment contains four bins of A3 values, followed by the second segment with four bins of PINV values. Each bin of a segment gives the value of the amplitude plus the values of each previous bin.

A normal cumulative curve shows an increasing negativity of A3 and an increasing activity (positivity) of the PINV. Both reach the same absolute value at the end.

From the normal cumulative curve, four major distinguishing types (A–D) can be constructed (**Figure 3**).

1. Type A: the absolute maximum value of A3 is clearly lower than the absolute maximum value of the PINV. The maximum value of the PINV is positive.
2. Type B: the absolute maximum value of the PINV is clearly lower than the absolute maximum value of A3. The maximum value of A3 is negative.
3. Type C: the absolute maximum value of A3 is clearly lower than the absolute maximum value of the PINV. The maximum value of the PINV is negative.
4. Type D: the absolute maximum value of the PINV is clearly lower than the absolute maximum value of A3. The maximum value of the A3 is positive.

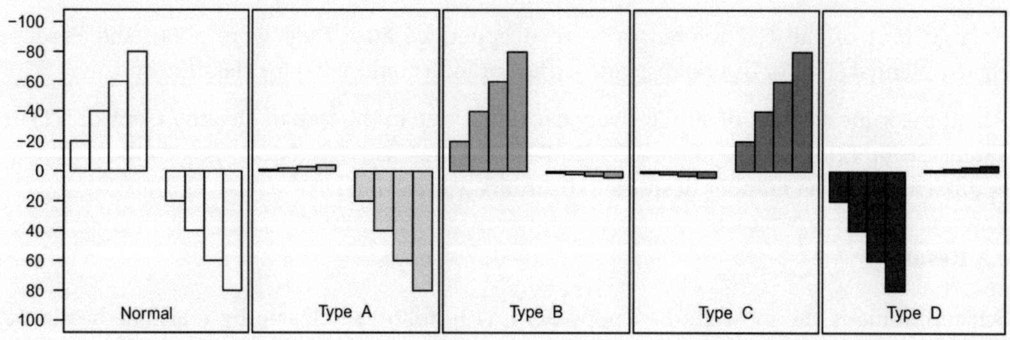

Figure 3. Normal and four abnormal types (A-D) of cumulative curve.

3. Literature

As mentioned before, as clinicians, we like to see research findings developed into clinical (diagnostic) tools. With this in mind, it seems useful to not only contrast current clinical findings of the CNV cumulative curve with research findings, but also assess the current status of

CNV research in regard to its potential as a clinical tool. This was done with a survey of literature and by applying the four-step approach for developing diagnostic tests in psychiatry [3].

3.1. Method

Relevant articles were selected on PubMed, from the National Center of Biotechnology Information (www.pubmed.gov), with the search formula "Contingent Negative Variation" [MAJR] NOT "mismatch negativity" AND (Humans[Mesh]).

This resulted in a total of 674 references which were scored following the proposed four-step model for developing diagnostic tests in psychiatry [3]:

Step 1 demonstrates the presence of a deviant variable in a target group compared to a healthy group and the test-retest reliability. This provides evidence of a consistent abnormality in the target group.

Step 2 demonstrates a significant differential prevalence of a deviant variable between disorders and a healthy group. This demonstrates the potential for clinical utility.

Step 3 demonstrates how promising the findings are for development as a diagnostic test by comparing the target group to proper control groups. It defines the test-performance characteristics and the clinical utility.

Step 4 extends the comparison of the target group versus control groups by using larger groups or by multicenter trials. This step demonstrates the clinical application and sets up standards for clinical use.

This procedure gives us a solid methodology to evaluate literature in prospect of the data analysis of the selected major disorder classes and clinical observations.

A large part of the 674 references were dropped because they were about the Bereitschaftspotential (194) or did not provide sufficient information (97) for classification.

About the same amount of studies were excluded, due to the lack of Healthy Control (71) or Target Group (219). The remaining 93 studies were then classified according to the four-step approach. Next, the number of studied disorders was counted.

3.2. Results

Schizophrenia is the most studied population (19), followed closely by migraine/headache (16). To a lesser extent, Parkinson's disease (7), depression (6) and dementia (5) were researched. The remaining disorders were mostly studied only once and are not considered here (**Table 1**).

The classification on the four-step approach method resulted in 84 studies classified at step 1 and 9 at step 2. In the studies reaching step 2, schizophrenia is the most prevalent disorder used as comparison group, seven out of the nine studies.

3.2.1. Schizophrenia

Studies show that there is a clear presence of a PINV in the schizophrenic patient. Half of the studies found a reduced amplitude of the CNV with an indication of a more frontal distribution.

Disorder	Number
Schizophrenia	19
Migraine/headache	16
Parkinson's disease	7
Depression	6
Dementia	5
Others	52
Total	105

Table 1. Number of disorders as comparison group in selected studies.

3.2.2. *Migraine/headache*

The migraine patient is characterized with increased amplitude, especially of the early CNV. Several studies also show a perturbed habituation.

Patients with headache rather show a PINV.

3.2.3. *Parkinson's disease*

In the Parkinson patient, both the early CNV and the late CNV are reduced in amplitude.

Two of the seven studies seem to indicate that there is a correlation between the amplitude and mental functioning.

3.2.4. *Depression and dementia*

It was not possible to come to any meaningful conclusion due to the use of a wide variety of paradigms, which made a comparison rather senseless.

3.3. Conclusions

From literature survey, it seems promising to use the CNV, at least for schizophrenia, since the studies demonstrate the potential for clinical utility (step 2). To a lesser extent, only two studies, depression also holds promise for clinical utility. However, due to a lack of consistent methodology across studies, further studies are needed to support this.

At present, the comparison of these studies suggests that it seems premature to link a deviant parameter to a single disorder: for example, low CNV amplitude can be found in a schizophrenic as well as a Parkinson patient.

Since there are more parameters (see Section 4.3) that can be measured on the CNV, it seems reasonable to use them in combination in an attempt to differentiate between disorders that show the same change on one parameter.

4. Clinical data analysis

4.1. Subjects

A total of 3757 clinical patients (mean ± SD age = 44.5 ± 15.9) were selected on the following criteria: hospitalized on the psychiatric ward, classified on the DSM-IV-TR and underwent a CNV.

After classifying the patients into the major DSM-IV categories, the groups that had a minimum of 100 individuals were retained.

This resulted in a total of 4196 clinical subjects assigned to six major disorder classes: (1) Substance-Related Disorders, (2) Mental Disorders Due to a General Medical Condition Not Elsewhere Classified, (3) Dissociative Disorders, (4) Schizophrenia and Other Psychotic Disorders, (5) Mood Disorders and (6) Impulse-Control Disorders Not Elsewhere Classified.

Due to the fact that a patient can have multiple DSM-IV numbers assigned, the total number of classified subjects is larger than the population total.

4.2. Results

4.2.1. A3

In all major categories, a minimum of 91% (mean ± SD = 92.2 ± 0.9) of the patients show a normal amplitude of the late CNV. There are also no major differences between the groups for increased or decreased amplitude (**Table 2**).

4.2.2. A5

At least 96% (mean ± SD = 97.5 ± 0.9) of the patients in each major group show a normal amplitude of the post-S2 amplitude. No significant differences were found between the groups for decreased or increased amplitude (**Table 2**).

4.2.3. PINV (A5 – A3)

A normal PINV amplitude was found for a minimum of 88% (mean ± SD = 89.4 ± 1.2) of the patients in each group. Up to 11.6% (mean ± SD = 10.5 ± 1.2) showed an increased amplitude of the PINV, but no significant differences were found between groups (**Table 2**).

4.2.4. Cumulative curve

The type B is the most prevalent with an average of 61.7% (±SD = ±3.0), with no significant differences between groups. Following is the type A with an average of 23.1% (±SD = ±2.9) and type C with an average of 12.2% (±SD = ±2.2). The smallest group is type D with an average of 3% (±SD = ±1.5). Once again, no significant differences between groups were found (**Table 3**).

DSM-IV-TR class		A3			A5			A5-A3		
	N	≤–2SD	≥2SD	NS	≤–2SD	≥2SD	NS	≤–2SD	≥2SD	NS
SRD	1525	0.6	8.2	91.2	1.2	1.7	97.0	8.9	0.4	90.8
MDG	766	0.4	6.7	93.0	0.9	2.3	96.7	11.5	0.4	88.1
DD	155	0.0	9.0	91.0	0.6	1.9	97.4	11.6	0.0	88.4
SPD	111	0.0	8.1	91.9	0.0	0.9	99.1	9.0	0.0	91.0
MD	1497	0.4	6.7	92.9	1.2	1.9	96.9	10.4	0.3	89.4
ICD	142	0.7	6.3	93.0	1.4	0.7	97.9	11.3	0.0	88.7
Average		0.4	7.5	92.2	0.9	1.6	97.5	10.5	0.2	89.4
SD		0.3	1.1	0.9	0.5	0.6	0.9	1.2	0.2	1.2

SRD: Substance-Related Disorders, MDG: Mental Disorders Due to a General Medical Condition Not Elsewhere Classified, DD: Dissociative Disorders, SPD: Schizophrenia and Other Psychotic Disorders, MD: Mood Disorders and ICD: Impulse-Control Disorders Not Elsewhere Classified.

Table 2. Percentage of subjects in each major diagnostic class that shows a significant decreased (≤–2SD), increased (≥2SD) or normal (NS) amplitude for each parameter.

4.3. Conclusions

Data analysis of clinical subjects show that for the six major class disorders studied, the amplitude of the late CNV and the presence or absence of a PINV is unable to differentiate between these classes. The cumulative curve can also not be used to differentiate between the studied major class disorders.

DSM-IV-TR class		Cumulative curve			
	N	Type A	Type B	Type C	Type D
SRD	1404	21.0	63.6	13.1	2.3
MDG	713	24.5	59.6	14.8	1.1
DD	140	26.4	58.6	9.3	5.7
SPD	105	19.0	66.7	11.4	2.9
MD	1394	22.2	60.6	14.3	2.9
ICD	133	25.6	60.9	10.5	3.0
Average		23.1	61.7	12.2	3.0
SD		2.9	3.0	2.2	1.5

SRD: Substance-Related Disorders, MDG: Mental Disorders Due to a General Medical Condition Not Elsewhere Classified, DD: Dissociative Disorders, SPD: Schizophrenia and Other Psychotic Disorders, MD: Mood Disorders and ICD: Impulse-Control Disorders Not Elsewhere Classified.

Table 3. Percentage distribution of the different types of cumulative curves in each major diagnostic class.

These negative results does not necessary mean that we need to abandon the CNV and the cumulative curve. First of all, there are more parameters than the two that have been used in this analysis:

- M1 (early CNV): average value between 500 and 700 ms after S1
- M2 (late CNV): average value of the 200 ms pre-S2 epoch
- M3 (post-sensory positivity): average value between 300 and 500 ms after S2
- M4 (post-motor positivity): average value between 300 and 500 ms after motor response
- M5 (post-sensory positivity): average value between 500 and 700 ms after S2
- M6 (post-motor positivity): average value between 500 and 700 ms after motor response
- SM1 (total surface between M1 and M2): sum of values of all points between 500 ms after S1 and start of S2 relative to the average baseline
- SM2 (total surface between 300 and 800 ms after S2): sum of values of all points between 300 and 800 ms after S2 relative to the average baseline

Apart from using a single parameter, any combination of parameters could be used to improve the differential diagnosis of disorders. This can be extended with a cumulative curve, comparing any pre-S2 averages with one of the post-S2 averages.

Secondly, the author's clinical experience has demonstrated that there is another potential for the CNV and the cumulative curve that has not been considered. Two clinical examples not only help to illustrate this, but also to give some indication why the data analysis was unable to differentiate between disorders.

5. Clinical examples

5.1. Bipolar affective disorder

This patient was examined five times on average every 139.25 days (±SD 40.66 days). The subject was initially examined with an indication of depression. Only on the second hospitalization the patient was sent for examination with the indication for bipolar affective disorder.

Two alternating types of cumulative curves were seen across the five subsequent visits: type A and type B (**Figure 4**). The type A corresponds to the depressive episode and type B matches the manic episode.

When the respective averages were made of the CNV curves for the depressive and manic episodes, classified as such using the cumulative curve (**Figure 5**), differences in amplitude of the late CNV (A3) and PINV (A5-A3) became apparent. Both episodes show decreased amplitude of the late CNV (A3) and also early CNV and is more pronounced in the depressive episode. The manic episode has a value of −9.572 μV for A3 and −0.207 μV for PINV and the depressive episode −3.961 μV for A3 and 8.254 μV for PINV (**Figure 6**).

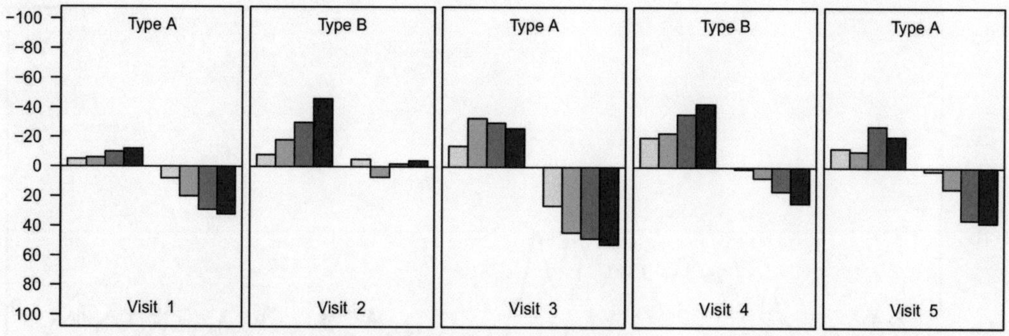

Figure 4. Cumulative curves of a bipolar affective disorder case for each visit.

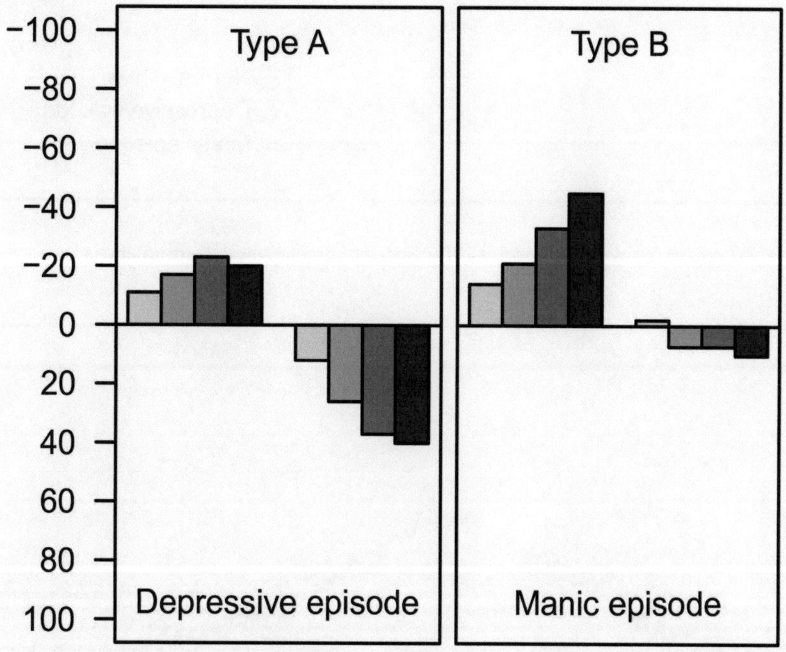

Figure 5. Cumulative curves of the averaged CNV curves of the depressive and manic episodes of a bipolar affective disorder case.

5.2. Stable depression with an episode of decline and recuperation period

This patient was examined 15 times on average every 228.21 days (±SD 75.20 days).

The subject is known with a recurrent major depressive disorder, which was stable as reflected in the cumulative curve of the CNV during eight subsequent visits, on average every 266.00 days (±SD 78.46 days), lasting six years. The stable period was interrupted with one episode of decline following a subsequent number of negative personal dramatic incidences. This was reflected in the CNV by an overall decline of the amplitude and a type C cumulative curve.

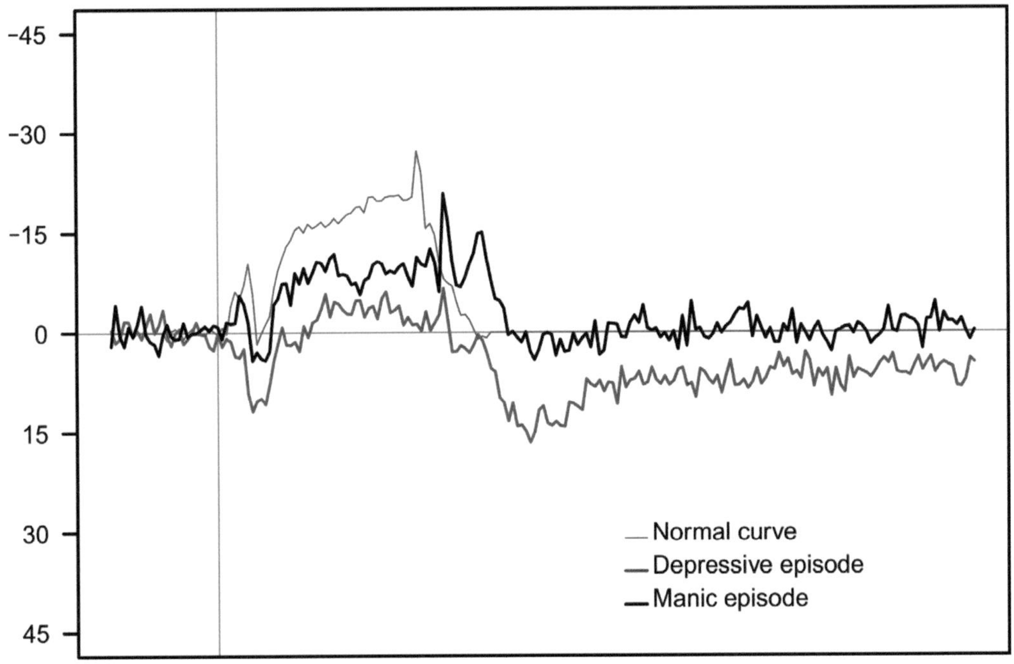

Figure 6. Average curves of the manic and depressive episode of a bipolar affective disorder case.

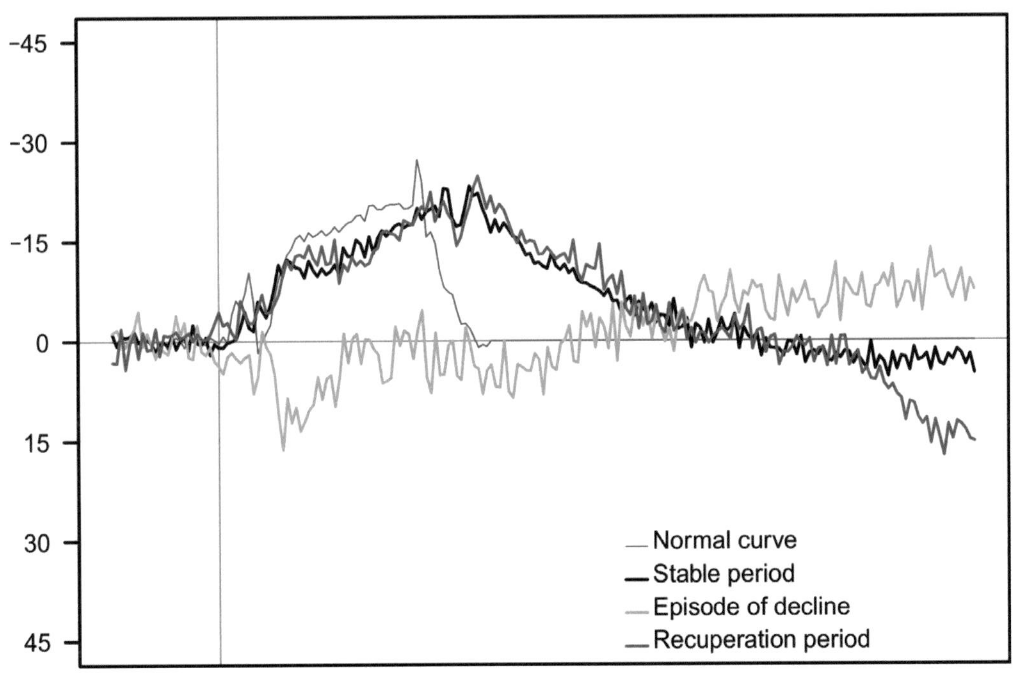

Figure 7. Average curves of the stable and recuperation period of a depressive case together with the curve of the single episode of decline.

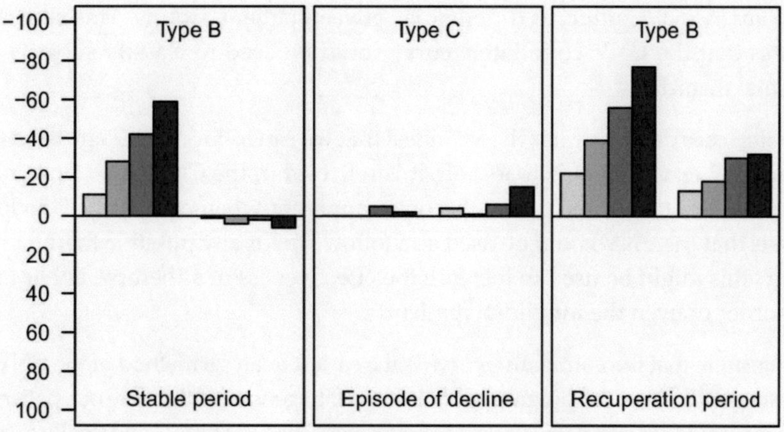

Figure 8. Cumulative curves of a depressive case: stable period, episode of decline and period of recuperation.

The patient recovered and corresponds to the average CNV of six subsequent visits, on average every 198.60 days (±SD 66.65 days). The overall amplitude returned to the previous stable period (**Figure 7**) and a type B cumulative curve was seen again (**Figure 8**).

However, the comparison of the cumulative curves of the recuperation and the stable period shows that the maximum amplitude of the PINV is higher in the last curve. From this, one could either conclude that the patient has not yet fully recovered or will continue to show a lasting impact of this decline (**Figure 8**).

6. Discussion

From literature survey, it seems that the CNV could only be applied for schizophrenics with certainty. Yet, one must remark that this is due to the lack of sufficient studies and the use of a standard paradigm within other patient populations.

From the data analysis of clinical patients, it is clear that the research findings for schizophrenics, the decreased amplitude and the presence of PINV are also found in other disorders. The data analysis also suggests that the CNV cannot be used for differential diagnosis.

Does this really mean that we need to discard the CNV from clinical practice?

To start answering this question, let's return to the two clinical examples.

From the bipolar case, we not only learn that the CNV can at least be used as an aid to diagnose bipolar affective disorder but also might explain why there is a lack of clinical useful findings in CNV research. Within the same patient group, each individual is not in the same state when examined in clinical practice or taking part in a scientific study. This averages out many or all differences one might find if we would take into account the current state of each subject. In mixing both episodes of the bipolar patient, we fail to see the difference in amplitude of the late CNV between the depressive and manic episodes. This mix might also explain the absence of significant differences between patient groups. This clinical case also demonstrates that the CNV cumulative curve could be used to classify subjects on current state of mental disorder.

The depression case demonstrates the strongest that we can follow the progression of a patient through time. When a patient is worse off, it is reflected in the CNV. Also, any change or no change at all in the state of the patient becomes apparent when comparing previous results. This suggests that the CNV could be used as a follow-up for any possible therapy that is initiated. Ideally, this might be used to increase the effectiveness of a therapy, either tailored to a specific disorder or even the individual patient.

Both cases illustrate that the cumulative curve is the easiest analysis method of the CNV to express the current state and follow the progression of a subject. In general, if the state of a patient continues to worsen, the cumulative curve will change from a type B, through type C and finally to type D in the most unfavorable situation (**Figure 9**). When the patient improves, the cumulative curve will change in the opposite direction. In contrast, the bipolar patient will swing back and forth between the types A and B, bipolar cycle (**Figure 9**), as long as there is no improvement in the patient's condition.

To return to the question, clinical observations seem to support the position that there is no need to abandon the use of the CNV in clinical practice. The CNV cumulative curve shows potential as a diagnostic aid, although limited at this time to bipolar affective disorder. However, the greatest potential might lie in the therapeutic decision process. When a therapy is initiated, its progress can be followed with the cumulative curve. From the proposed model, one can expect that the success of a therapy will be reflected in the cumulative curve. If successful, the natural progression would deviate from its natural course and develop toward a normal cumulative curve. If this can be substantiated, the CNV would have a role as an aid to increase therapeutic effectiveness and reducing treatment cost.

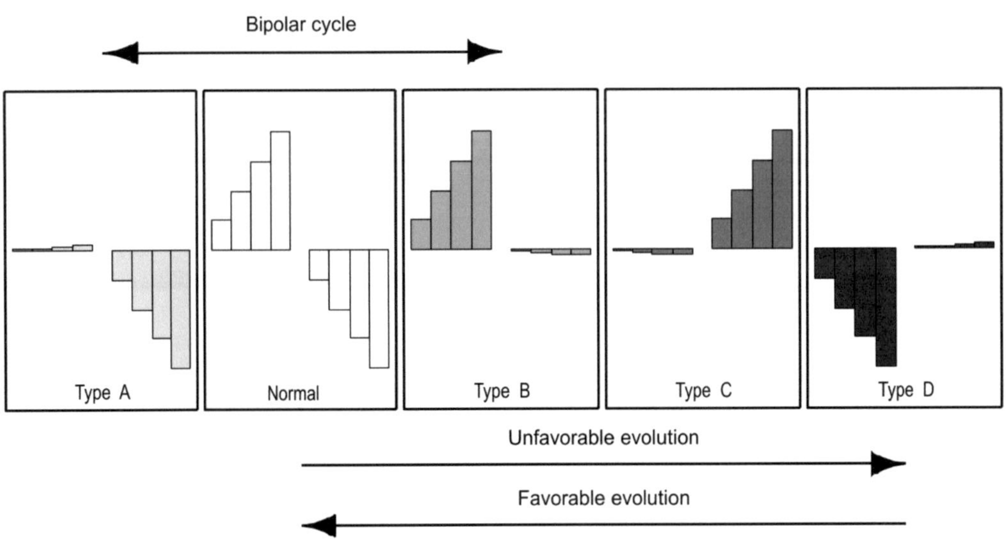

Figure 9. Evolution cycles of the cumulative curve.

It must be clear that more research is needed to confirm these clinical observations. Apart from this, recommendations for further research have been highlighted in this chapter.

First, research should be aware that there is a need for clinical (diagnostic) tools in psychiatry. The four-step approach can be used to assess at what level a test is and ascertain if more research is needed at the current level or the requirements are met to move forward into the next level.

More and better interaction is needed between researchers and clinicians to guide the development process. This close collaboration could result in a full development of a test, from discovery of a consistent abnormality to a standardized clinical (diagnostic) tool. It can also assure that a test will closely meet the needs of the clinician.

We should also not forget previous research. Technology has further developed and offers many more possibilities. This should not blind us from the more 'simple' analysis that has been done in the past. Instead of moving these aside, we should use them to look at them from a new perspective. With current technology, we are no longer restricted to select one or a few parameters, but can also do multiple combinations with a simple push of a button.

Author details

Daniel Dumalin

Address all correspondence to: daniel.dumalin@azsintjan.be

AZ Sint-Jan Brugge-Oostende AV, Ostend, Belgium

References

[1] Walter WG, Cooper R, Aldridge VJ, McCallum WC, Winter AL. Contingent negative variation: An electric sign of sensori-motor association and expectancy in the human brain. Nature. 1964;**203**(4943):380-384. DOI: 10.1038/203380a0

[2] McCallum WC. Human slow potential research: A review. In: McCallum WC, Curry S, editors. Slow Potential Changes in the Human Brain. New York: Plenum Press; 1993. pp. 1-12

[3] Boutros N, Fraenkel L, Feingold A. A four-step approach for developing diagnostic tests in psychiatry: EEG in ADHD as a test case. The Journal of Neuropsychiatry and Clinical Neurosciences. 2005;**17**(4):455-464. DOI: 10.1176/jnp.17.4.455

[4] Timsit-Berthier M, Delaunoy J, Koninckx N, Rousseau J. Slow potential changes in psychiatry. I. Contingent negative variation. Electroencephalography and Clinical Neurophysiology. 1973;**35**(4):355-361. DOI: 10.1016/0013-4694(73)90191-0

[5] Timsit-Berthier M, Delaunoy J, Gerono A. Morphological analyses of the CNV in psychiatry: Comparison of resolution mode and cumulative curve methods. In: Otto D, editor.

Multidisciplinary Perspectives in Event-Related Brain Potential Research: Proceedings of the Fourth International Congress on Event-Related Slow Potentials of the Brain (EPIC IV). Washington: Environmental Protection Agency; 1978. pp. 389-391

[6] Timsit-Berthier M, Gerono A. Manuel d'interprétation des potentiels évoqués endogènes (P300 et VCN). Bruxelles: Éditions Mardaga; 1998. p. 64

[7] Tecce JJ. Contingent negative variation (CNV) and psychological processes in man. Psychological Bulletin. 1972;**77**(2):73-108. DOI: 10.1037/h0032177

Chapter 5

The Visual Evoked Potential in Idiopathic Inflammatory Demyelinating Diseases

Silvio Pessanha Neto, Luiz Carlos Pinto and
Regina Maria Papais Alvarenga

Additional information is available at the end of the chapter

http://dx.doi.org/10.5772/intechopen.69311

Abstract

Within the group of inflammatory idiopathic demyelinating diseases, there is a great number of diseases that have an initial attack in common, including visual. Multiple sclerosis (MS) is a chronic, demyelinating, immune-mediated disease, with considerably varying prevalence and incidence. Neuromyelitis optica (NMO), which until recently was considered a variant of MS, is currently considered an independent entity. However, it resembles MS, because it is an immune-mediated disease characterized by the simultaneous or sequential involvement in time of optic neuritis and extensive demyelinating myelitis. Fifty percent of patients with MS have isolated optic neuritis. However, the frequency of abnormalities ranges from 57 to 100% in visual evoked potential (VEP). Several studies have evaluated the clinical, evolutive, and demographic characteristics of idiopathic optic neuritis and demonstrated their differences among the cases related to MS and NMO. The most common changes in VEP studies in multiple sclerosis are as follows: increased interocular differential latency of P100 wave and the absolute increase in latency of P100 wave. New studies indicate that VEP pattern in NOM spectrum syndromes is different from that of MS.

Keywords: neuromyelitis optica, visual evoked potential, optic neuritis, multiple sclerosis, P100 wave

1. Introduction

Visual evoked potential (VEP), which has been known for more than 40 years, consists of electric signals mainly generated by the occipital cortex in response to visual stimulation. They are generally used to assist in the identification of pathologies that impair the visual pathways in any of its segments.

INTECH
open science | open minds

© 2017 The Author(s). Licensee InTech. This chapter is distributed under the terms of the Creative Commons Attribution License (http://creativecommons.org/licenses/by/3.0), which permits unrestricted use, distribution, and reproduction in any medium, provided the original work is properly cited. (cc) BY

2. Technical aspects

2.1. Stimulation

In VEP studies, two types of stimulation, photostimulators and pattern reversal, are normally used. The most used photostimulators are the common photographic flash and goggles, which are used for swimming, or similar ones, in which each lens is replaced by a cloudy plate, with minute light bulbs installed in its internal face, for the luminous extremities to face each eye. These stimulators are connected and synchronized with the equipment system of stimulation, scanning, and promediation. Flash stimulation is not only the oldest one, used in clinical neurophysiology since the 40s, but also the most common method used during many years and until the beginning of the 70s to obtain VEP. The flash stimulates almost all the retina, mainly the peripheral one, because of its capacity to capture the variations of room luminosity. Depolarization waves provoked by stimuli radiate mainly to the pretectal zones and thalamic nuclei, ending up in lesser amount in the visual cortex, and in larger amount in the association areas, on the nonvisual cortex. When the flash is used, it is necessary that the contralateral eye is very well occluded, because the capture of potentials generated by the stimulation of the unexamined side can lead to false-negative results.

One of the main disadvantages in the responses obtained with this technique is its great variability, both among individuals and in the same patient; it can also vary according to the electroencephalographic activity at that moment. Therefore, flash stimulation was gradually replaced by more reliable methods, being reserved for the cases of patients with very low visual acuity, encephalopathies, and other pathologies.

In the case of goggles, the light bulbs or light emitting diodes (LEDs) on them are generally red, and each one has 3 mm of diameter. At each series of stimuli, the stimulator is adjusted so that one side of the goggles remains with the light off, occluding vision in one of the patient's eyes, while in the stimulated side the LEDs are on and are turned off with a frequency of 1–2 Hz. In each side, at least two series of 60–260 stimuli are applied.

Both flash and goggles stimulate the entire retina and provide results that are more qualitative than quantitative. However, they are very useful, because they can stimulate the retina even when the eyes are closed, as it occurs in the case of patients in coma, children or adults who do not want to cooperate [1, 2].

Currently, despite the utility of the photostimulators, the most commonly used visual stimulation is the pattern reversal that, as already mentioned, consists of a monitor in the screen of which there is an image with black and white squares, similar to a chessboard. During stimulation, the squares alternate their colors. The white ones become black, and the black ones become white, successively, at a frequency of 1–2 Hz. These constant inversions stimulate the macular area of the retina, more specifically the foveal zone, radiating to the lateral geniculate body, and then to the primary visual cortex, in area 17. However, for this, it is important that the patient keeps the look fixed on the center of the monitor screen [3].

In the stimulation by the reversal pattern, the patient should remain seated in front of the monitor, with the eyes keeping a distance of 70 cm to 1 m from the screen. As the stimulation is monocular, the contralateral eye is kept occluded and at least two series of 60–260 stimuli in each side of the visual pathway are applied.

The distance between the patient's eye and the monitor, related to the size of the each reversal pattern square side, provides the visual angle used in the test. Despite the existence of tables, the simplest way to calculate the visual angle is to multiply the width of the square by 3450 and divide the result by the distance in millimeters between the screen and the patient's eyes. This way, the visual angle is obtained in minutes. To convert this unit to degrees, we simply divide the resulting value by 60 [4–6].

The visual angle used is very important, because it influences in the exam results. In adults, visual angles with 10–20 arcmin produce responses of higher amplitude. The visual angles of more than 15 min stimulate mainly the fovea, and those with more than 40 min stimulate more the parafoveal retina. The central visual field is the most responsible for P100 amplitude, because the central vision has greater cortical representation than the peripheral vision. Problems in the central vision can change P100 amplitude without modifying its latency. In the diagnostic routine, angles of 28–32 arcmin are more frequently used.

The equipment visual stimulation module allows the use of a reversal pattern, the squares of which can vary in size according to the examiner interest. In practice, they have options of squares of 0.5, 1.0, 2.0, or 5 cm of side. With this, depending on the clinical case and the patient's visual acuity, the series of stimuli can be repeated with the use of different visual angles.

There is a direct relation between the visual angle and the exam result. The smaller the squares, the more reliable the responses. However, in the patients with visual acuity deficit, the use of bigger squares is necessary, because even losing some precision, it is crucial that the patient visualizes and identifies the reversal pattern; otherwise, it will not be possible to capture responses, or falsely abnormal responses will be obtained [7–9]. Before undergoing a VEP study, the patient should have an ophthalmologic evaluation for measuring his/her visual acuity, aiming to rule out refractive defects [10].

The reversal pattern can be one of "full field," which is the most used, or of "half field." In the half-field pattern, the chessboard image fills only one of the halves of the monitor screen, with the other remaining entirely black. This allows selective stimulation of the nasal or temporal field of the retinas, serving both for the diagnosis and to inform if the lesion is prechiasmatic, chiasmatic, or retrochiasmatic [11].

Some authors have used colored reversal patterns, but the results have not been advantageous. There is also the "bar grating reversal pattern," known as gratings, which is little used and on which, instead of the squares, the monitor presents horizontal or vertical, white and black bars, which alternate colors consecutively too [12, 13].

In brief, because of its higher precision and sensitivity, the reversal pattern is the most commonly used. The flash and goggles are used in cases where it is supposed that the patient is not looking at the center of the screen, in comatose patients, in surgical monitoring, or in children who have difficulty fixing attention on the monitor.

2.2. Capture

The most common assembly consists of the colocation of an active electrode in Oz, on the occipital cortex, with the reference in Fz, and the ground in the frontal area in Fpz. However, depending on the number of channels available, other assemblies can be made. In a second channel, for example, the active on Oz can be used, with the reference on Pz or Cz. Although being a capture type that is more used in half-field stimulations, the active electrodes can be fixed on points located 5 and 10 cm lateral to Oz, in known leads, such as R5, R10, L5, or L10, in which the letters "R" and "L" represent the right side and the left side, respectively.

The visual response is formed by three waves, which form the M-like image. The first deflection is N75, which is negative. It is followed by a positive, sharper, and deeper wave, with a mean latency of 100 ms, that it is the P100. The third wave, N145, is also negative as the first one. However, the most important response is P100, because it is the most defined and the one that has greater reproducibility in normal people. It represents the occipital cortex depolarization to the applied stimuli. The normal values for each one of these responses can vary in accordance with the laboratory and its equipment.

The first negative potential (N1) has a latency between 60 and 90 ms; the first positive potential (P100) has a latency of 85–120 ms, with an average value of 100 ms. The second negative potential (N2) varies from 125 to 155 ms.

The amplitude of the P100 response varies between 3 and 21 μV. The amplitude of the N1 and N2 responses is a little lower and usually measures less than 12 μV for N1 and less than 16 μV for N2. N1 and N2 responses vary a lot, and in some cases, they are not easy to be identified. In part, for this reason, maximum attention is usually given to P100 response, which is easy to obtain and reproduce in normal people, and has well-defined parameters.

P100 latency and the parameters of normality vary from a laboratory to another due to the technique used, the type of equipment, and the age of the examined people. In laboratories where a population of young patients predominates, the mean and maximum latencies accepted as normal will be lower than those obtained in laboratories where older patients predominate.

The ideal is that each examining physician determines his/her own normal values. For this, at least 30 normal and representative people should be examined. The results obtained should be analyzed, and a mean latency should be determined for the whole group. The normal parameters are obtained, adding two or three standard deviations to this average value. If this is not possible, then normal values published by authors of high credibility in the area can be used in a more practical way.

As previously cited, the inconsistency and variability of the N1 and N2 potentials discouraged most authors, and all the attention and importance was given to P100 response. Despite all this, Pavot believes that, although the N1 and N2 parameters are wider than those of P100,

http://dx.doi.org/10.5772/intechopen.69311

there are certain limits of normality. According to his experience, in certain pathological conditions, as, for example, in multiple sclerosis (MS), in some cases the manifestations appear with an increase of N2 latency, or with its disappearance [14].

In the interpretation of the exams, absolute P100 latency is valued, as well as the difference between P100 latencies in one side and the other (interocular latency) and also its amplitude, which is compared with that of the contralateral eye. A difference of amplitude higher than 50% between the two sides is significant.

2.3. Factors that interfere with the evoked visual responses

When performing VEP studies, it is extremely important to give much attention to some details that can affect the results. In addition to age, sex, visual acuity, and other physiological factors, the visual evoked potentials are also affected by technical parameters, such as electrode positioning; luminosity in the examination room; level of concentration of the patient, who has to keep the look persistently directed to the center of the screen of the stimulation monitor; visual angle; luminosity and brightness of the monitor screen; and type of stimulator; filters.

In reversal pattern stimulation, variations in the quality of the stimuli, such as brightness, luminosity, frequency, and size of the square, can affect the responses, changing the morphology, amplitude, and latencies of its components.

Brightness reduction increases latency and reduces P100 amplitude, in 15 ms and 18%, respectively, for each log unit that reduces luminosity. The reduction of the contrast, which can be calculated in percentage according to the difference of luminosity between the white and black squares, also affects P100, increasing its latency and reducing its amplitude.

As previously mentioned, the size of the square determines the visual angle used and has direct influence in the responses. The use of smaller squares makes the study more accurate, but in turn, its use will depend on the level of visual acuity of the patient, since they stimulate the fovea, and the refractive defects produce out-of-focus retinal images, increasing P100 latency. The larger squares with up to 5 cm of side are less affected by the visual acuity variations.

The electrodes should be well located, because positioning them on points that are out of those normally proposed result in absent, or dispersed, responses with potentials of modified shapes, low amplitude, and unreal latencies. Loose, badly fixed electrodes produce artifacts that make the identification of the potentials difficult.

Noise excess in the room can affect the patient's concentration capacity in the examination, making the look to deviate from the stimulator. During the examination, the examiner should ensure that the patient is kept awake all the time, and with the look focused on the center of the monitor screen. If the patient closes the eyes or deviates the look from the screen, the potential amplitude can lower and the latencies can increase, leading to false-positive results.

In cases of restless patients, with concentration difficulty, as it is common to occur in children and even in those cases when the examiner suspects that the patient is purposefully trying to interfere with the examination, a stimulation performed with special glasses or goggles is recommended.

Some authors recommend to avoid exam performance in the evening because at this time the patient is frequently tired, sleepy and it is more difficult to remain concentrated, and with the look fixed on the screen.

Special attention should be given to refractive errors, and to retinal diseases. Patients who wear glasses or contact lenses have to be alerted to take them on the day of the examination, because decreased visual acuity slightly reduces amplitude of the responses, and can also change their latency.

Age modifies P100 latency, following a variable curve that is descending in the two first decades of life, steady until the fifth or sixth decade, and crescent above 60 years. Women tend to have little shorter latencies than men. However, menopausal women can present a higher P100 latency than men in a similar age.

In healthy individuals, there is no evidence of P100 alterations caused by the same increase of temperature or even by exercises. However, the exercises can reduce P100 amplitude in multiple sclerosis patients with impairment of the visual pathways [15].

3. Criteria for the analysis of the visual evoked potentials

In the analysis of the visual responses obtained, the examiner should value the following parameters: absolute latency; interocular latency difference, or interocular differential latency; P100 response amplitude; interocular amplitude difference or interocular differential amplitude; and potential morphology.

3.1. Absolute latency

The examiner evaluates whether the latency of the potential obtained is within the normal range, that is, the normal average with more or less three standard deviations. A latency that exceeds the limits of normality, with the possibilities of technical errors discarded, indicates a defect of sensory conduction in the studied visual pathway.

3.2. Interocular differential latency

The parameter for this measure should be determined in each laboratory, but in general it varies from 5 to 8 ms. An interocular differential latency above these values is usually associated with pathologies. Many times this latency increase is the first manifestation of some diseases. In many cases of optic neuritis, the interocular latency change can be the only abnormality detected.

3.3. P100 response amplitude

P100 amplitude can vary in normal people, but an amplitude of 1–5 μV, or the absence of P100 generally means a pathological condition [16].

http://dx.doi.org/10.5772/intechopen.69311

3.4. Interocular amplitude difference

There is much controversy on the limit of normality for the value of the amplitude difference between the responses obtained in the two eyes. When the stimulation for the "full-field reversal pattern" is used, some authors advocate that, to be abnormal, there must be an amplitude reduction of at least 80%, or total absence of P100. However, for most authors, a reduction of the differential amplitude between 50 and 75% or more indicates visual pathway impairment. In the stimulation of the hemifield, there is a consensus that an amplitude difference of more than 50% is abnormal.

3.5. Morphology

In a normal person, the P100 response has the format of a letter "V." An alteration in the potential shape or its disappearance represents an abnormality [16].

4. Clinical applications

Any pathology, regardless of its nature, which affects the ocular structures responsible for the reception of light and images, the visual pathway, or the cortex, can lead to changes in the visual evoked responses. In the pure retinal pathologies, as it occurs in the retinitis pigmentosa, the important changes in the ophthalmologic examination and in the electroretinogram are very striking and easy to be identified. On the other hand, it is important to remember that retinal responses take a long time to disappear in cases of brain death.

In the diseases affecting the receptors and the visual pathways, or the visual pathways and the cortex, frames with coincident or divergent findings can be observed in the VEP studies. The frames known as coincident are those in which no difference in the type of response obtained is observed, either with the use of the flash, or with the reversal pattern, that is, the responses are normal or abnormal in the two methods. In the frames with divergent findings, the neurophysiological abnormalities are manifested in only one of the techniques, with normal responses being observed in the other.

The visual evoked potentials are used in the investigation of neurological impairment, mainly in the suspicion of optic neuritis, multiple sclerosis, and compressive lesions affecting the visual pathways. Diseases affecting the optic nerve or causing its demyelination are the more frequently associated with changes of visual response latency. To understand the neurophysiological diagnosis, and the alterations caused by pathologies throughout the optic pathways, it is necessary to know these pathways anatomy [17, 18].

The visual pathways are formed from the retina, through the chaining of three types of neurons. The first receptor neurons, which are the cones and the rods, make synapsis with bipolar cells, and these ones with a third type of neuron, which are the ganglion cells. The ganglion cell axon junction forms the optic nerves, which take the visual impulses to the lateral geniculate bodies in the diencephalon, where they make synapses with neurons that go to the occipital cortex through the geniculocalcarine tract.

In its route to the geniculate bodies, the optic nerve fibers from the nasal portion of each retina cross to the opposite side on the level of the optic chiasm. Thus, from the chiasm, each optic tract consists of the optic fibers from the temporal retina on the same side and of the fibers that were formed in the contralateral nasal retina. This peculiarity of the optic pathways to cross part of their fibers in the chiasm, continuing as a mixed tract that has fibers of the retinal portion of the ipsilateral eye, and also of the contralateral eye, causes the most varied visual syndromes, depending on the location of the lesion.

5. CNS idiopathic inflammatory demyelinating diseases

Demyelinating optic neuritis represents the most frequent cause of transitory visual loss in young adults, affecting 2.6 men and 7.5 women per year, for each 100,000 inhabitants. The average age for its occurrence is 31 years. The optic pathway demyelination causes blocks or delays in the visual pathways conduction, with consequent alterations in the studies of visual evoked potentials. Demyelination precedes the inflammatory process, which is the real responsible for the reversibility of the picture. Thus, inflammation improvement contributes for the rapid visual improvement after a crisis. In addition, mainly in young people, remyelination is another important factor in the recovery of vision. However, when an axonal lesion and a more persistent demyelination occur, improvement is not usually complete.

Visual loss in optic neuritis can be preceded in some days by ocular pain in the affected side, which tends to resolve. This pain is possibly caused by the tension on the inflamed nerve. About 70% of the adults initially present with a unilateral picture, but in 30% it can affect both eyes. The visual loss can be sudden, progressing in a few hours, or can have a slower progression, taking some days to be installed. In 7% of the cases, this time is of 1–2 weeks.

The diagnosis of optic neuritis is one of exclusion, what makes the investigation of other diseases affecting the optic nerve, such as hereditary, metabolic, toxic, vascular, and compressive diseases, indispensable.

In the isolated demyelinating optic neuritis, the magnetic resonance reveals changes on the affected nerve in 84% of the cases. In 34% of the patients, the exam also shows changes in the asymptomatic side. In addition, 50–70% of these patients show multifocal demyelinating lesions in the corpus callosum and on the periventricular white substance or in other parts of the encephalon.

Within the group of the idiopathic inflammatory demyelinating diseases, there is a great number of pathologies that can have an initial outbreak in common, either visual, motor, sensitive, proprioceptive, cerebellar, medullary or of the brainstem, characterizing the so-called clinically isolated syndrome—CIS. It is only after the second outbreak and evaluation of the complementary exams that it is possible to establish or suggest the definitive diagnosis, such as MS. This aspect can sometimes confuse physicians who attend the patient and delay the treatment.

MS diagnosis is based on the identification, at history taking, of two acute episodes with a duration of at least 24 h, and evidence at the neurological examination of objective signs of functional system impairment, indicating inflammatory lesions located on different topographies in the CNS.

It was only in 1983 that complementary methods, such as magnetic resonance and evoked potentials, were introduced in the MS diagnosis criteria proposed by Poser, with the purpose to identify subclinic inflammatory lesions. In the last decade, MRI was used to confirm temporal and spatial dissemination of inflammatory lesions in the neuroaxis and then to anticipate the clinical diagnosis of MS. Currently, with a patient with monofocal or multifocal clinically isolated syndrome, the dissemination in the space can be demonstrated in the MRI through T2 hyperintense lesions, in at least two of the four regions of the CNS: periventricular, juxtacortical, infratentorial, and spinal. Temporal dissemination is proven by the presence of a new T2 lesion, or a contrast-enhancing lesion, when serial MRI scans are compared, or by the coexistence of asymptomatic contrast-enhancing or nonenhancing lesions in a single initial examination. These radiological criteria should only be applied to young patients, with strong clinical suspicion of MS characterized by the presence of clinical signs of acute CNS impairment, which presents with outbreaks, having a suggestive behavior of an inflammatory disease, and after ruling out all the diseases secondarily affecting the white substance. The application of these criteria aims to anticipate the clinical diagnosis and, consequently, the beginning of the treatment, since all the FDA-approved medicines from 1993 on for MS act in the initial phase of the disease, reducing inflammation and the annual rate of outbreaks [20].

MS is a chronic disease, with greatly variable prevalence and incidence, dependent on ethnicity and demographic region, with the highest indices being described in Caucasian populations living in regions of the North hemisphere, places of cold weather. In its more prevalent clinical form, there are outbreaks and remissions that affect individuals from 20 to 40 years of age, and with predominance in women. Currently, it is estimated that more than 300,000 Americans have a definite diagnosis of the disease and, because it affects young patients in full activity, it has strong sociocultural impact [21].

NOM, which was included among MS variants, is currently considered an independent condition. The historical description of this disease was a clinical, and anatomical and pathological case report in 1894, in France, by Eugene Devic. This condition was called Devic's disease, and during a century, the diagnosis was based on the identification of an acute, monophasic inflammatory disease, characterized by severe and bilateral ON, and transverse myelitis (TM), installed simultaneously or in a short interval of time. Only after the 1990s studies of independent series published in different western and eastern populations started to describe recurrent cases of NOM, where the index events occurred separately for a variable period of time, and were followed by new acute episodes, affecting the spinal cord and the optic nerve [22–24]. Only recently the presence of lesions in locations other than the spinal optic axis was accepted, with inflammatory lesions being demonstrated, although in lower frequency, in the brainstem and encephalon [23]. In Asian individuals, the selective and severe involvement of the optic nerves and spinal cord is very typical. In this region, this syndrome is classified as a variant of MS, defining two distinct subtypes: the opticospinal form (OSMS), which has characteristics that are similar to the remittent-recurrent form of western NOM, and the conventional form, which is similar to the classic MS, as it is described in western patients [25].

Currently, many works in the literature try to define the probability of an isolated and initial case of idiopathic optic neurite to progress, to keep its monophasic course, to be associated with myelitis or with outbreaks that impair other areas of the CNS. Among several significant aspects, the main justifications for these studies regarding the progression risk are related to the patient's prognosis, and the possibility to search early therapies at the first signs of isolated idiopathic ON, since this can be the first symptom, not only of NOM, but also of MS.

The clinical and evolutive characteristics of idiopathic ON have been analyzed by the Optic Neuritis Study Group (1995). The most relevant predictive factor of MS development after 15 years was the presence of changes in the brain MRI in the occasion of a visual outbreak [26]. Other works tried to identify the main characteristics of the cases of NOM-related ON, evidencing that in this condition the visual involvement is generally more severe and bilateral [24].

In 2007, Wingerchuk et al. [27], from the Mayo Clinic, defined a group of conditions that were catalogued as syndromes of the NOM spectrum. These entities, although independent, are elements of the same group. They are as follows: neuromyelitis optica (NOM), extensive idiopathic myelitis, recurrent monocular optic neuritis (rON) or simultaneously bilateral optic neuritis (BON), Asian-type opticospinal multiple sclerosis (OSMS), optic neuritis, or extensive idiopathic myelitis associated with systemic autoimmune disease, optic neuritis, or idiopathic myelitis associated with typical brain lesions of NOM (hypothalamus, corpus callosum, and periventricular region) or lesions in the brainstem.

6. VEP in idiopathic inflammatory demyelinating diseases

Since 1972, the pattern of visual response abnormalities of patients with ON, obtained with this method, has been studied [28], with the MS-related VEP characteristics of the optic neuritis being well-defined. However, there are few articles in the literature discussing the VEP pattern in NOM. Currently, MS pattern is frequently used to analyze the VEP of patients with NOM, configuring an interpretation bias.

The most common changes in the VEP studies in MS, in descending order of prevalence and importance, are as follows: the absolute increase of P100 wave latency, this wave morphology changes, and finally the absence of response. Frederiksen and Petrera [29] followed patients in the acute phase of ON since the beginning of the symptoms, with performance of VEP, repeating it in subsequent months. Among these patients, 35.5% had the definite diagnosis of MS, and had, according to the authors, a significant relation with delays of P100 wave latency in the VEP. They also observed subclinical changes of the optic nerve, with the presence of abnormal VEP in asymptomatic eyes.

Several studies emphasize VEP sensitivity in the evaluation of ON in patients with demyelinating diseases, even overcoming the optical coherence tomography, as Naismith et al. showed in 2009 [30]. At that time, the authors evaluated patients with different diagnosis related to the demyelinating diseases (CIS, MS, and NOM) that had at least one episode of ON in the last six months and tried to compare VEP sensitivity to the optical coherence tomography, in the evaluation of clinical and subclinical ON. VEP showed higher sensitivity both in the evaluation of clinically identified ON (81% vs. 60%, $p = 0.002$), and in the cases of evaluation of asymptomatic eyes.

In the study by Matthews et al. [29], a total of 223 individuals were evaluated, with 186 with a diagnosis of MS and 37 healthy controls. In this study, the sample was submitted to VEP, to evaluate the differences in the pattern of responses to this exam, considering the groups of definite, probable, possible MS, and the control group. It showed that while all controls had VEP responses within normal range, the group with definite MS had a higher percentage of change when compared to the other groups. It was also observed that the main characteristic of the VEP that was changed was P100 wave latency in these patients.

Frederiksen and Petrera [31] followed patients in the acute phase of ON since the beginning of symptoms, with performance of VEP, and its repetition in subsequent months. Among these patients, 35.5% had definite diagnosis of MS, and had, according to the authors, a significant relation with delays of P100 wave latency in the VEP. They also observed subclinical changes of the optic nerve, with the presence of abnormal VEP in asymptomatic eyes.

In order to enhance the sensitivity for the detection of optic nerve affections with the use of VEP, and to analyze its effectiveness in the early evaluation of patients with suggestive pictures of ON and MS, Davidson et al. in 2004 [32] examined 124 individuals with VEP using two different intensities of contrast. Although they observed a higher number of abnormal VEP with reduced contrast, when they followed the outcome they noticed that they were false positives.

Considering that NOM is a pathology that is different from MS, and its strong relation with the presence of anti-AQP4 serum antibodies, Watanabe et al., in 2009 [33], decided to study the abnormalities of the findings of evoked potentials of patients with definite diagnosis of MS, and the presence of positivity for anti-QP4. In this study, they observe that in the anti-QP4-negative MS group the delay in the latency of the P100 wave is evident, following the literature. However, the anti-QP4-positive group presents a higher percentage of patients with absence of this wave, revealing a more severe lesion ($p = 0.003$). They also conclude that the antibody positivity, and the absence of response, showed significant relation with a more serious progression of visual impairment (OR = 35.432%, $p < 0.001$).

Naismith et al., in 2009, evaluated patients with different diagnoses related to demyelinating diseases (CIS, MS, and NOM) that had at least one episode of ON in the last 6 months and tried to compare VEP sensitivity with the optic coherence tomography in the evaluation of clinical and subclinical ON. VEP showed higher sensitivity both in the evaluation of clinically identified ON (81% vs. 60%, $p = 0.002$) and in the cases of evaluation of asymptomatic eyes (75% vs. <20%) [34].

Neto et al., in 2013, conducted a study where 19 patients, with NOM diagnosis, with 74% being Afro-Brazilians, underwent VEP study. Of the 38 eyes examined, 18 (47.37%) showed no visual evocable response. Of the 20 eyes (52.63%) where VEP responses were detected, 18 (90%) had P100 wave latency within normal range, while only 2 (10%) had increase of the latency of this wave. Regarding P100 wave amplitude, 11 of the 20 eyes (65%) that generated visual responses had values below that considered normal in the study. Seven (35%) had amplitudes ≥ 5.8 μV, being considered normal. In 65% of the 20 eyes where the visual response was evocable, a reduction of the P100 wave amplitude was found, with normal latency [34].

Ringelstein et al., in 2014, reproduced the study by Neto et al., and analyzed the medical records of 43 Caucasian patients with NOM diagnosis, and compared the findings of their VEP with those of 81 healthy patients. The authors find reduced amplitude in 12.3%, long latencies in 41.9%, and absence of response in 14% of NOM eyes, suggesting that VEP in NOM would have a heterogeneous standard. However, the frequencies of amplitude reduction and the absence of response are greater than those observed in all the studies of patients with MS. In the article, they suggest that the difference in sample results, compared to the study by Neto et al. could be explained by ethnic issues that distinguish the populations studied [19]. In addition, in the study by Ringelstein et al., the reference value for amplitude normality is lower than that used by Neto et al., being 3.0 and 5.8 μV, respectively [35].

In 2015, Chirapapaisan et al. evaluated hospital medical records of patients being investigated for MS, with no definite diagnosis. VEP was analyzed, along with the confirmation of the subsequent diagnosis of MS through McDonald diagnostic criteria (2005). Twelve of the 35 patients (34%) converted to MS, and 23 (66%) did not have diagnostic confirmation. P100 Latencies and differences of interocular latency were longer in the clinically definitive MS (CDMS) than in non-CDMS patients (p = 0.002, 0.001, respectively). All patients of the group that converted to MS had P100 latencies higher than 102ms, the average of the patients with no diagnosis of MS, thus providing 100% of sensitivity. No patient developed MS with P100 latency <102 ms. Brain lesions in the magnetic resonance were significantly associated with CDMS development (p = 0.001). Therefore, the previsibility to develop CDMS was higher when the P100 latency delay and the brain lesions of magnetic resonance were concomitantly present [36].

In the last decade, recurrent ON, not associated with MS and NOM, was classified according to its clinical presentation in recurrent isolated form (RION) and chronic recurrent (CRION) form [37]. In CRION and RION, ON is more severe than in MS, leading to severe, bilateral visual impairment that can cause amaurosis. However, while a high prevalence of anti-AQP4 antibody positivity is observed in NOM, in CRION 95% of the patients present negativity for anti-AQP4 [38]. In all syndromes cited, the OCT reveals significant reduction of the fiber layers of the temporal and nasal retina.

7. Conclusions

We can conclude that VEP has recently been sufficiently studied and shows differences between the classic pattern of MS and that of NOM, where the responses are more heterogeneous, and the reduction of P100 wave amplitude and the absence of response are more prevalent [34], but data in the literature are lacking about the VEP pattern in RION, CRION, syndromes of the NOM spectrum, and OSMS of MS.

http://dx.doi.org/10.5772/intechopen.69311

Author details

Silvio Pessanha Neto[1,2,3*], Luiz Carlos Pinto[3] and Regina Maria Papais Alvarenga[2]

*Address all correspondence to: silvio.neto@estacio.br

1 Estacio de Sa University, Rio de Janeiro, Brazil

2 Federal University of the State of Rio de Janeiro, Rio de Janeiro, Brazil

3 Luiz Carlos Pinto Clinical Neurophysiology Center, Rio de Janeiro, Brazil

References

[1] Kraut MA, Arezzo JC, Vaughan FIG. Intracorticalgenerators of the flash VEP in monkeys. Electroencephalography and Clinical Neurophysiology. 1985;**62**:300-312

[2] Korth M. Pattern evoked responses and luminanceevoked responses in human electroretinogram. Journal of Physiology. 1983;**337**:451-469

[3] Halliday AM, McDonald WI, Mushin J. Visual evoked response in diagnosis of multiple sclerosis. British Medical Journal. 1973;**4**:661-664

[4] Celesia CG. Steady state and transient visual evoked potentials in clinical practice. Annals of the New York Academy of Sciences. 1982;**388**:290

[5] TECA Corp. Visual angles. In: TECA Corp., editor. Tab 26.83. New York; 1981

[6] Harter MR, White CT. Evoked cortical responses to check-board patterns: Effects of check-size as a function of visual acuity. Electroencephalography and Clinical Neurophysiology. 1970;**28**:48-54

[7] Hennerici M, Wenzel D, Freund HJ. The comparison of small-size rectangle and checkerboard stimulation for the evaluation of delayed visual evoked responses in patients suspected of multiple sclerosis. Brain. 1977;**100**:119-136

[8] Neima D, Regan D. Pattern visual evoked potentials and spatial vision in retrobulbar neuritis and multiple sclerosis. Archives of Neurology. 1984;**41**:198-201

[9] Oken BS, Chiappa KH, Gil E. Normal temporal variability of P100 latency. Electroence phalography and Clinical Neurophysiology. 1987;**68**:153-156

[10] Rowe MJ III. The clinical utility of half field pattern reversal visual evoked potential testing. Electroencephalography and Clinical Neurophysiology. 1982;**53**:73-77

[11] Parker DM, Salzen EA. Latency changes in human visual evoked response to sinusoidal gratings. Vision Research. 1977;**17**:1201-1204

[12] Skrandies W. Scalp potential fields evoked by grating stimuli: Effect of spatial, frequency and orientation. Electroencephalography and Clinical Neurophysiology. 1984;**58**:325-332

[13] Chiappa KH. Principles of evoked potentials. In: Chiappa KH, editor. Evoked Potentials in Clinical Medicine. 2nd ed. New York: Raven Press; 1990. pp. 1-35

[14] Stockard JJ, Hughes IF, Sharbrough FW. Visualevoked potentials to electronic pattern reversal: Latency variations with gender, age, and technical factors. American Journal of EEG Technology. 1979;**19**:171-204

[15] Lana MA. Neurite óptica. In: Tilbery CP, editor. Esclerose Múltipla no Brasil. São Paulo: Editora Atheneu; 2005. pp. 129-148

[16] Chiappa KH. Evoked Potentials in Clinical Medicine. 3rd ed. Philadelphia: Lippincott-Raven; 1997

[17] Halliday AM, McDonald WI, Mushin J. Delayed visual evoked responses in optic neuritis. Lancet. 1973a;**1**:982-985

[18] Shahroskhi F, Chiappa KH, Young RR. Pattern shift visual evoked responses in 200 patients with optic neuritis and/or multiple sclerosis. Archives of Neurology. 1978;**35**:65-71

[19] Halliday AM, McDonald WI, Mushin J. Delayed pattern evoked responses in optic neuritis in relation to visual acuity. Transactions of the Ophthalmological Societies of the United Kingdom. 1973;**93**:315-324

[20] Miller DH, Johnson G, McDonald WI, DuBoulay EPGH, Kendall BE, Moseley IF. Detection of optic nerve lesions in optic neuritis with magnetic resonance imaging. Lancet. 1986;**1**:1490-1491

[21] Poser CM. The epidemiology of multiple sclerosis: A general overview. Annals of Neurology [S.l.]. 1994;**36**(Suppl 2):180-193

[22] O'Riordan J, Gallagher L, Thompson J, Howard R, Kingsley D, Thompson E, Mcdonald W, Miller D. Clinical, CSF, and MRI findings in Devic's neuromyelitis optica. Journal of Neurology, Neurosurgery, and Psychiatry. 1996;**60**:382-387

[23] Pittock SJ, Weinshenker BG, Wingerchuk D, Lucchinetti CF. Autoimmune neurological accompaniments of neuromyelitis optica (NMO). Annals of Neurology. 2006;**60**:S41

[24] Papais-Alvarenga RM, Carellos SC, Alvarenga MP, Holander C, Bichara RP, Thuler LCS. Clinical course of optic neuritis in patients with relapsing neuromyelitis optica. Archives of Ophthalmology [S.l.]. 2008;**126**(1):12-16

[25] Kira J, Kanai T, Nishimura Y, Yamashaki K, Tobimatsu S, Kobayashi T. Western versus Asian types of multiple sclerosis. Immunogenetically and clinically distinct disorders. Annals of Neurology. 1996;**40**:569-574

[26] Optic Neuritis Study Group. Multiple sclerosis risk after optic neuritis: Final optic neuritis treatment trial follow-up. Archives of Neurology [S.l.]. 2008;**65**(6):727-732

[27] Wingerchuk DM, Lennon VA, Lucchinetti CF, Pittock SJ, Weinshenker BG. The spectrum of neuromyelitis optica. Lancet Neurology. 2007;**6**(9):805-815

[28] Halliday AM, Mcdonald WI, Mushin J. Delayed visual evoked response in optic neuritis. Lancet [S.l.]. 1972;**1**(7758):982-985

[29] Frederiksen JL, Petrera J. Serial visual evoked potentials in 90 untreated patients with acute optic neuritis. Survey of Ophthalmology. 1999;**44**:S54-S62

[30] Naismith RT, Tutlam NT, Xu J, Shepherd JB, Klawiter EC, Song S, Cross AH. Optical coherence tomography is less sensitive than visual potentials in optic neuritis. Neurology. 2009;**73**:46-52

[31] Matthews WB, Small DG, Small M, Pountney E. Pattern reversal evoked visual potential in the diagnosis of multiple sclerosis. Journal of Neurology, Neurosurgery, and Psychiatry. 1977;**40**:1009-1014

[32] Davidson AW, Scott RF, Mitchell KW. The effect of contrast reduction on pattern-reversal VEPs in suspected multiple sclerosis and optic neuritis. Documenta Ophthalmologica. 2004;**109**:157-161

[33] Watanabe A, Matsushita T, Doi H, Matsuoka T, Shigeto H, Isobe N, Kawano Y, Tobimatsu S, Kira J. Multimodality-evoked potential study of anti-aquaporin-4 antibody-positive and -negative multiple sclerosis patients. Journal of the Neurological Sciences. 2009;**281**:34-40

[34] Neto SP, Alvarenga RM, Vasconcelos CC, Alvarenga MP, Pinto LC, Pinto VL. Evaluation of pattern-reversal visual evoked potential in patients with neuromyelitis óptica. Mult Scler. 2013 Feb;**19**(2):173-8

[35] Ringelstein M, Kleiter I, Ayzenberg I, Borisow N, Paul F, Ruprecht K, Kraemer M, Cohn E, Wildemann B, Jarius S, Hartung HP, Aktas O, Albrecht P. Visual evoked potentials in neuromyelitis optica and its spectrum disorders. Multiple Sclerosis. 2014;**20**(Suppl 5):617-620. DOI: 10.1177/1352458513503053

[36] Chirapapaisan N, Laotaweerungsawat S, Chuenkongkaew W, Samsen P, Ruangvaravate N, Thuangtong A, Chanvarapha N. Diagnostic value of visual evoked potentials for clinical diagnosis of multiple sclerosis. Documenta Ophthalmologica. 2015;**130**(Suppl 1):25-30. DOI: 10.1007/s10633-014-9466-6

[37] Petzold A, Plant GT. Diagnosis and classification of autoimmune optic neuropathy. Autoimmunity Reviews. 2014;**13**(4-5):539-545

[38] Petzold A, Pittock S, Lennon V, Maggiore C, Weinshenker BG, Plant GT. Neuromyelitis optica-IgG (aquaporin-4) autoantibodies in immune mediated optic neuritis. Journal of Neurology, Neurosurgery, and Psychiatry. 2010;**81**:109-111